# Processing of
# Seismic Reflection Data
# Using MATLAB™

# Synthesis Lectures on Signal Processing

Editor
**José Moura**, *Carnegie Mellon University*

Synthesis Lectures in Signal Processing will publish 50- to 100-page books on topics of interest to signal processing engineers and researchers. The Lectures exploit in detail a focused topic. They can be at different levels of exposition—from a basic introductory tutorial to an advanced monograph—depending on the subject and the goals of the author. Over time, the Lectures will provide a comprehensive treatment of signal processing. Because of its format, the Lectures will also provide current coverage of signal processing, and existing Lectures will be updated by authors when justified.

Lectures in Signal Processing are open to all relevant areas in signal processing. They will cover theory and theoretical methods, algorithms, performance analysis, and applications. Some Lectures will provide a new look at a well established area or problem, while others will venture into a brand new topic in signal processing. By careful reviewing the manuscripts we will strive for quality both in the Lectures' contents and exposition.

Processing of Seismic Reflection Data Using MATLAB™
Wail A. Mousa and Abdullatif A. Al-Shuhail
2011

Fixed-Point Signal Processing
Wayne T. Padgett and David V. Anderson
2009

Advanced Radar Detection Schemes Under Mismatched Signal Models
Francesco Bandiera, Danilo Orlando, and Giuseppe Ricci
2009

DSP for MATLAB™ and LabVIEW™ IV: LMS Adaptive Filtering
Forester W. Isen
2009

DSP for MATLAB™ and LabVIEW™ III: Digital Filter Design
Forester W. Isen
2008

DSP for MATLAB™ and LabVIEW™ II: Discrete Frequency Transforms
Forester W. Isen
2008

DSP for MATLAB™ and LabVIEW™ I: Fundamentals of Discrete Signal Processing
Forester W. Isen
2008

The Theory of Linear Prediction
P. P. Vaidyanathan
2007

Nonlinear Source Separation
Luis B. Almeida
2006

Spectral Analysis of Signals: The Missing Data Case
Yanwei Wang, Jian Li, and Petre Stoica
2006

Processing of Seismic Reflection Data Using MATLAB™
Wail A. Mousa and Abdullatif A. Al-Shuhail

ISBN: 978-3-031-01406-2    paperback
ISBN: 978-3-031-02534-1    ebook

DOI 10.1007/978-3-031-02534-1

A Publication in the Springer series
SYNTHESIS LECTURES ON SIGNAL PROCESSING

Lecture #10
Series Editor: José Moura, *Carnegie Mellon University*
Series ISSN
Synthesis Lectures on Signal Processing
Print 1932-1236    Electronic 1932-1694

# Processing of Seismic Reflection Data Using MATLAB™

Wail A. Mousa and Abdullatif A. Al-Shuhail
King Fahd University of Petroleum & Minerals

*SYNTHESIS LECTURES ON SIGNAL PROCESSING #10*

# ABSTRACT

This short book is for students, professors and professionals interested in signal processing of seismic data using MATLAB. The step-by-step demo of the full reflection seismic data processing workflow using a complete real seismic data set places itself as a very useful feature of the book. This is especially true when students are performing their projects, and when professors and researchers are testing their new developed algorithms in MATLAB for processing seismic data. The book provides the basic seismic and signal processing theory required for each chapter and shows how to process the data from raw field records to a final image of the subsurface all using MATLAB. The MATLAB codes and seismic data can be downloaded at `http://www.morganclaypool.com/page/mousa`.

# KEYWORDS

seismic data, amplitude correction, seismic noise attenuation, spiking deconvolution, static correction, migration

*To my dearly beloved father, mother, wife, and children,* W.M.

*To my dear wife and children,* A.S.

# Contents

# Preface

This book aims to provide students and interested experts from the DSP community with the necessary industrial background of seismic data processing using MATLAB, starting from loading the data, displaying, quality control of the data until obtaining a seismic image. We decided to write it after offering and teaching various related courses, for several years, in both the Electrical Engineering and Earth Sciences Departments at King Fahd University of Petroleum & Minerals (KFUPM). We hope that the book will motivate students to conduct research in multidisciplinary areas and assist DSP experts in further developing advanced algorithms for processing seismic data. We have written most of the book MATLAB's functions and m-files and used few MATLAB functions from the available GNU seismic MATLAB codes provided by the Signal Analysis and Imaging Group, Department of Physics, University of Alberta, Canada. Although this book is written to be self contained, there exist many useful MATLAB tutorials on the internet in addition to the Mathworks documentation for MATLAB, which will benefit the interested reader in understanding the written m-files and M-functions. Nevertheless, as far as this book is concerned, it is sufficient for the reader to know the basics of using and programming in MATLAB. The readers may enjoy the seismic data and the book MATLAB codes by downloading them following the link `http://www.morganclaypool.com/page/mousa`.

In Chapter 1, we provide the reader with an overview of seismic data processing. Then, in Chapter 2, we examine the two-dimensional (2-D) real seismic data set and extract various useful processing information and data for the proceeding chapters. Chapter 3, some commonly used quality control techniques in order to correct for amplitude losses in real seismic data sets. Chapter 4 explains what the main signals are that we are looking for and how to analyze the data in the spectrum domain. Frequency filtering to attenuate ground roll noise contaminating our real data is also explained in the same chapter. In Chapter 5, we present the seismic convolution model and how we can apply deconvolution to enhance the vertical resolution of the data. After that, in Chapter 6, we carry the seismic data processing forward by sorting the data from shot gathers into common mid-point gathers, picking seismic stacking velocities, applying normal moveout (NMO) correction and, finally, stacking the data. To further enhance the seismic data, we then apply in Chapter 7 residual static correction to the NMO-corrected data and we stack the data again. We then apply seismic migration to enhance the horizontal resolution of our data in Chapter 8. Finally, we close with some concluding remarks in Chapter 9.

Wail A. Mousa and Abdullatif A. Al-Shuhail
September 2011

# Acknowledgments

We would like to thank KFUPM for its continuous support of our teaching and research efforts, where special thanks are due to H.E. Dr. K. Al-Sultan, KFUPM Rector, for his constant encouragements and support. Also, we would like to thank Prof. Lina Karam, School of Electrical, Computer, and Energy Engineering, Arizona State University, for encouraging us to write this book. Special thanks are due to Ayman Al-Lehyani, Lecturer, Earth Sciences Department, KFUPM, for his assistance in fixing some problems with the provided reflection seismic data headers. As well, we thank Professor Joel S. Watkins, Professor Emeritus of Texas A&M University for allowing us to use the data set provided with this book. Finally, we thank Joel Claypool for his encouragement to complete this work.


Wail A. Mousa and Abdullatif A. Al-Shuhail
September 2011

CHAPTER 1

# Seismic Data Processing: A Quick Overview

## 1.1 INTRODUCTION

Among many geophysical surveying techniques, seismic reflection is the most widely used and well-known geophysical technique. Seismic reflection data can be processed to reveal details of geological structures on scales from the top tens of meters of the Earth's crust to its inner core [1, 2]. Part of its success lies in the fact that the raw seismic data is processed to produce seismic sections which are images of the subsurface structure. A geologist can then make an informed interpretation by understanding how the reflection method is used and seismic sections are created. The analysis of seismic data is performed for many applications such as petroleum exploration, determination of the earth's core structure, monitoring earthquakes, etc. [2, 3].

Seismic signals are generated by a source (transmitter), such as an explosion, which then propagate through earth layers. Some of these signals will be reflected, refracted and lost due to attenuation). At the surface, the reflected signals are then recorded by a receiver. The strength of the reflected signal depends on the impedance contrast between adjacent layers. In summary, a seismic survey analysis scenario involves collection of data by an array of receivers (geophones for land and hydrophones for marine), transmission over a narrow band channel, and storage of data for processing, and interpretation [1, 2, 4].

A seismic trace represents a combined response of a layered ground and a recording system to a seismic source wavelet. Any display of a collection of one or more seismic traces is termed a *seismogram*. Assuming that the pulse shape remains unchanged as it propagates through such a layered ground, the resultant seismic trace may be regarded as the convolution of the input impulse with a time series known as a *reflectivity function*, which is composed of spikes (delta functions). Each spike has an amplitude related to the reflection coefficient at a layer boundary and a traveltime[1] equivalent to the two-way reflection time from the surface to that boundary. Furthermore, the reflection time series represents the impulse response of the layered ground, which is basically the output for a spike input. Since the source wavelet has a finite length, individual reflections from closely-spaced

---

[1]Traveltime is the time difference between zero time and the arrival time of a seismic event. It can be a one-way time such as for direct waves or two-way time such as for reflected waves.

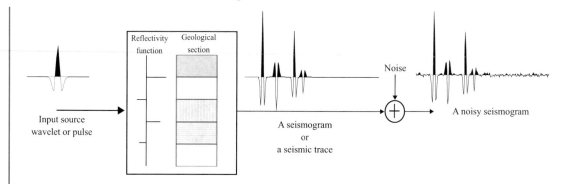

**Figure 1.1:** Convolution seismic data model. A seismic pulse is convolved with the reflection coefficient log (reflectivity function) to get a seismic trace. The reflection coefficient log is related to the geological section of the subsurface through the reflection coefficient of each geological boundary and the two-way travel time.

boundaries may overlap in time on the resultant seismogram (seismic section). Figure 1.1 represents a typical seismic convolution model [1].

Due to many factors, unwanted waves such as surface waves, corrupt the seismic records with noise (unwanted energy). This unwanted energy includes random or incoherent noise such as instrument signals and coherent noise like the ground roll noise. A typical example for a noisy seismic section is shown in Figure 1.2. As a consequence, of the above effects, seismic traces generally have a complex appearance and reflection events are often not recognized without the application of suitable processing techniques. The purpose of processing such data can, in general, be viewed as the process of attenuating noise and then determining the input pulse and removing that to give the reflectivity function, which ultimately allows the determination of the acoustic impedances (or related properties) of the subsurface layers.

## 1.2    SEISMIC DATA PROCESSING

Seismic data processing can be considered as a sequence of cascaded operations that attenuate/remove noise accompanying seismic data as well as making geometrical corrections such that the final image will truly show a map (seismic image) of the subsurface. Processing of seismic data includes, but is not limited to: filtering, common mid-point (CMP) sorting, velocity analysis, normal move-out (NMO) correction, and stacking. Each seismic trace has three primary geometrical factors which determine its nature: shot position, receiver position, and the position of the subsurface reflection point. The last factor is the most critical [1]. Before processing, this position is unknown but a good approximation can be made by assuming that this reflection point lies vertically under the position on the surface mid-way between the shot and the receiver for that particular trace. This point is referred to as *common mid-point* or *common depth-point* (CDP). Traces reflected from

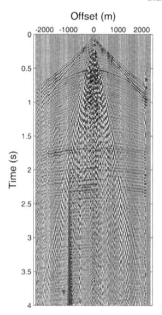

**Figure 1.2:** A typical example for a seismic section (http://cwp.mines.edu). The horizontal axis represents the offset of each seismic receiver (recorder) from the source where each records a trace with respect to the two-way travel time (vertical axis). Clearly, this section contains various noise types.

the same CMP define a CMP gather. The CMP gather is important for seismic data processing because the subsurface velocity can be derived using it. In general, the reflection seismic energy is very weak and it is essential to increase the signal-to-noise ratio (SNR) of most data. Once the velocity is known, the traces in CMP gathers can be corrected for NMO, which is basically a way of correcting for time differences which occur due to offset[2] in a CMP gather, i.e., to get the equivalent of a *zero-offset* trace. This implies that all traces will have the same reflected pulses at the same time, but with different random and coherent noise. So combining all the traces in a CMP gather will average out noise and the SNR increases. This process is known as stacking.

In general, the main objectives of seismic data processing are to improve the seismic resolution and increase the SNR of the data. These objectives are achieved through three primary stages. In their usual order of application, they are:

1. deconvolution, which increases the vertical resolution;.

2. stacking, which increases the SNR; and

[2]An offset is the distance from the source-point to a geophone or the center of a geophone group.

3. migration, which increases the horizontal resolution.

In addition to these primary stages, secondary processes may be implemented at certain stages to condition the data and improve the performance of these three processes. Figure 1.3 shows a conventional seismic data processing flow [2]. In the following section, we elaborate on the processing steps in this figure.

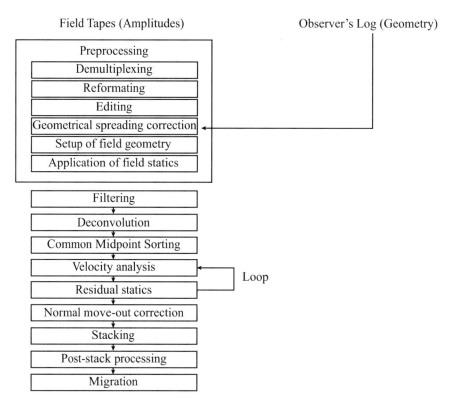

Figure 1.3: A conventional seismic data processing flow (after [2]).

1. Preprocessing: This process involves a series of steps to condition the data and prepare it for further quality control and processing including:

   • demultiplexing

   • reformatting

   • trace editing

   • gain application

   • setup of field geometry

- application of field statics

2. Filtering is used to attenuate components of the seismic signals based on some measurable property. It is an important step in order to proceed further with the other seismic data processing steps that will help geophysicists to better analyze and interpret the acquired data.

3. Deconvolution is performed along the time axis to increase vertical resolution by compressing the source wavelet to approximately a spike and attenuating noise and unwanted coherent energy such as multi-path signals.

4. CMP sorting transforms the data from shot-receiver (shot gather) to midpoint-offset (CMP gather) coordinates using the field geometry information.

5. Velocity analysis is performed on selected CMP gathers to estimate the stacking, root-mean squared (RMS), or NMO velocities to each reflector. Velocities are interpolated between the analyzed CMPs.

6. Residual static correction is usually needed for most land data. It corrects for lateral variations in the velocity and thickness of the weathering layer.

7. NMO correction and muting: The stacking velocities are used to flatten the reflections in each CMP gather (NMO correction). Muting zeros out the parts of NMO-corrected traces that have been excessively stretched due to NMO correction.

8. Stacking: The NMO-corrected and muted traces in each CMP gather are summed over the offset (stacked) to produce a single trace. Stacking M traces in a CMP increases the SNR of this CMP by $\sqrt{M}$.

9. Poststack processing includes time-variant band-pass filtering, dip filtering, and other processes to enhance the stacked section.

10. Migration: Dipping reflections are moved to their true subsurface positions and diffractions are collapsed by migrating the stacked section.

In the remaining parts of this book, we are going to practically explore most of the above processing steps along with application to a typical real seismic data set. Note that the above steps are general and depending on the data type, land or marine, the accompanying noise types, and/or the acquisition conditions, the above processing steps may vary as well.

CHAPTER 2

# Examination of A Real Seismic Data Set

## 2.1    INTRODUCTION

When geophysicists meet with their seismic data for the first time, they carefully must look into the seismic data itself and its header, which contains information and tabulation of parameters used to acquire the data. Usually, the data is stored in magnetic tapes or hard disks and is saved in various standard data formats such as the well-known SEG-Y or Seismic Unix formats [2, 5]. We have already read the seismic data (that we are going to process in this book) and its header information and stored them in a MATLAB data format file called `Book_Seismic_Data.mat`. The aim of this chapter is, therefore, to examine this data along with its header information using MATLAB. This will assist us in further analysis and processing of this real reflection seismic data in the remaining chapters of the book.

## 2.2    DESCRIPTION OF THE SEISMIC REFLECTION REAL DATA SET

The real data set we will use to illustrate the processing codes consists of a two-dimensional (2-D) land line from east Texas, USA. The following are some important parameters about the data:

- Number of shots = 18.

- Source type = dynamite in 80-100-ft depth holes.

- Number of channels per shot = 33.

- Receiver type = Vertical-component geophones.

- Array type = 12-element inline.

- Number of traces in line = 594.

- Receiver interval = 220 ft.

- Shot interval is variable.

- Time sampling interval = 2 milliseconds (ms)

- Number of time samples per trace = 1501.

- Data format = SEG-Y.

- Byte swap type = Big endian.

- Data file name = data.sgy.

- Geometry has already been set up and recorded in the trace headers.

- Uphole times at shot locations have been recorded in the trace headers.

- An 8-64-Hz bandpass filter has been applied to the data in the field.

## 2.3   EXAMINING THE DATA SET

Let us now examine the 2-D seismic data set and its header information by loading the file
Book_Seismic_Data.m. By loading the data in the MATLAB workspace we are going to find
two variables as follows:

```
1  >> load Book_Seismic_Data.mat
2  >> whos
3    Name            Size                 Bytes   Class        Attributes
4
5    D               1501x594           7132752   double
6    H                  1x594           3303008   struct
```

### 2.3.1   HEADER INFORMATION

After loading the data, we will find that variable H is a $1 \times 594$ structure array with many fields such
as: the time sampling interval dt, the number of time samples per trace ns, the offset offset, the
trace numbers tracl, the shot gather numbers fldr, etc. To access any one of these fields, say for
example, the offset and store its values in a new vector called offset, then:

```
1  >> offset=[H.offset];
2  >> whos
3    Name            Size                 Bytes   Class        Attributes
4
5    D               1501x594           7132752   double
6    H                  1x594           3303008   struct
7    offset             1x594              4752   double
```

The reader can explore more fields in a similar way. We have written a MATLAB function called
extracting_geometry.m which can be used to get useful geometrical information such as the
source coordinates $s_x$ and $s_y$, the receiver coordinates $g_x$ and $g_y$, the source elevations $g_z$, etc. The

following MATLAB code shows an example of how one can extract certain geometrical acquisition parameters:

```
1  load('Book_Seismic_Data.mat','H')
2  [sx,sy,gx,gy,shot_gathers,num_trace_per_sg,sz,gz]=
3  extracting_geometry(H);
```

These are different geometrical variables obtained from the seismic header structure H. We then can use such variables in the following code:

```
1  figure,stem(shot_gathers,num_trace_per_sg)
2  xlabel('Shot gather numbers','FontSize',14)
3  ylabel('Number of traces/shot gather','FontSize',14)
4  axis([0,max(shot_gathers)+1,0,max(num_trace_per_sg)+2])
5  set(gca,'YMinorGrid','on')
```

Also, here is another example:

```
1  figure,plot(sx,'.')
2  xlabel('Number of traces','FontSize',14)
3  ylabel('Sources x-axis locations (ft)','FontSize',14)
4  axis tight
5  grid
```

Figure 2.1 shows different useful plots all of which are obtained from the seismic header variable H. A final example is plotting the so-called seismic stacking chart (see Figure 2.2) using the MATLAB function stack_chart.m where the vertical axis shows the shot number and the horizontal axis shows the shot and receiver x-coordinate (we will discuss stacking charts in more details in chapter 6):

```
1  num_shots=length(shot_gathers);
2  stacking_chart(sx,gx,num_shots,num_trace_per_sg);
```

The reader may enjoy more plots when running the m-files written for this chapter.

## 2.3.2   DISPLAYING SEISMIC DATA

There exist several ways of displaying seismic sections such as seismic shot gathers and CMP gathers. The most commonly used displays are the following.

1. The wiggle display: plots seismic trace amplitudes as a function of time.

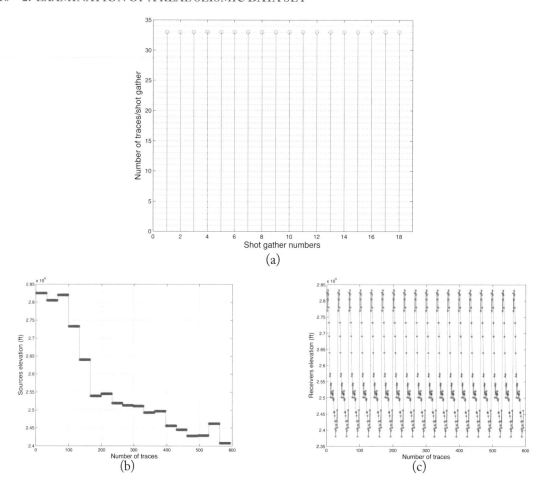

**Figure 2.1:** Various geometrical information plots: (a) the number of traces per shot, (b) the source elevation profile for each trace and (c) the receiver elevation per seismic trace.

2. The variable area display: shades the area under the wiggle trace to make coherent seismic events[1] evident.

3. The variable density display: represents amplitude values by the intensity of shades of gray (and sometimes in colors).

---

[1]A seismic event is the arrival of a new seismic wave, usually indicated by a phase change and an increase in amplitude on a seismic record. It may be a reflection, refraction, diffraction, surface wave, or random noise.

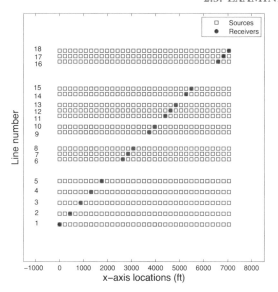

**Figure 2.2:** The stacking chart plot of the seismic data.

The following MATLAB script uses the written function `extracting_shots.m` to extract, for example, shot gather number 8. Then this shot gather is displayed using, respectively, the variable area display scheme and variable density display scheme in both gray and colors (see Figure 2.3):

```
1  load Book_Seismic_Data.mat
2  shot_num=8;
3  p=0;
4  [Dshot,dt,dx,t,offset]=extracting_shots(D,H,shot_num,p);
5  scale=1;
6  mwigb(Dshot,scale,offset,t)
7  xlabel('Offset(ft)','FontSize',14)
8  ylabel('Time(s)','FontSize',14)
9  figure,simage_display(Dshot,offset,t,0)
10 xlabel('Offset(ft)','FontSize',14)
11 ylabel('Time(s)','FontSize',14)
12 figure,simage_display(Dshot,offset,t,1)
13 xlabel('Offset(ft)','FontSize',14)
14 ylabel('Time(s)','FontSize',14)
```

Also, if we display shot gather number 16 (see Figure 2.4), we notice that trace number 31 amplitudes are increasing as time increases. This may require editing by muting this trace since it will affect the subsequent processing steps. In chapter 3, we shed more light on this issue. The reader may as well be interested in displaying a group of seismic shot gathers concatenated together using the same function `extracting_shots.m`. For example, one can extract shot gathers number 4-6

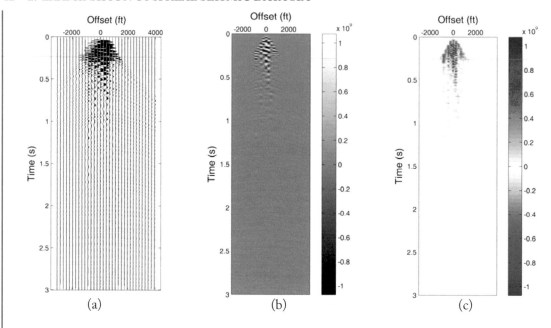

**Figure 2.3:** Various displays for seismic data shot gather number 8: (a) variable area display, (b) gray-scaled variable density display and (c) colored variable density display. The color bars in (b) and (c) refers to the amplitude dynamic range of the data.

where he/she must provide `shot_num=4:6;` in line 6 of the above script. Figure 2.5 shows these extracted shot gathers in their variable area display. The provided m-files illustrates more examples.

## 2.4    THE PROPOSED PROCESSING WORKFLOW

Based on the examination of the seismic data and its header, we are now ready to perform the following processing steps on the data:

1. Preprocessing involving only the gain application using various methods

2. Ground roll removal via bandpass filtering

3. Spiking deconvolution

4. CMP sorting

5. Velocity analysis on several CMPs using the velocity spectrum method

6. Residual static correction using the surface-consistent method

7. NMO correction and muting

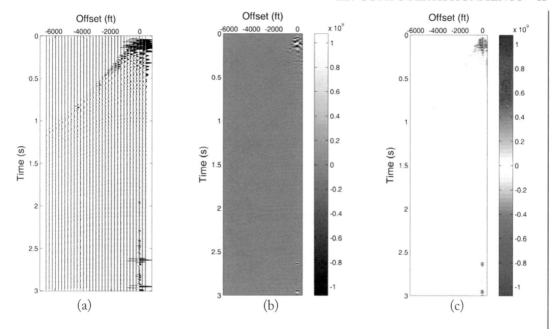

**Figure 2.4:** Various displays for seismic data similar to those in Figure 2.3 but for shot gather number 16. Clearly, trace number 31 (at offset 0) requires muting (replacing it by zeros) as will be discussed later in chapter 3.

8. Stacking

9. Migration using Stolt (F-K) post-stack time migration

## 2.5    COMPUTER ASSIGNMENTS

1. Load the seismic data Book_Seismic_Data.mat.

2. Examine the seismic data header information and obtain the following information: common-depth-point (CDP) values, number of time samples $nt$, spatial sampling interval $dx$ and time sampling interval $dt$.

3. Display shot gathers 12–15 (as the case in Figure 2.5) using the wiggle plotting, gray-scaled image plotting, colored image plotting, and wiggle on top of colored plotting. Comment on the shot gather amplitudes. What do you notice from a trace to another?

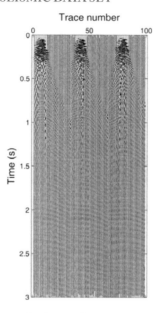

**Figure 2.5:** Variable area display of seismic shot gathers 4-6.

## 2.6  USEFUL MATLAB FUNCTIONS

load, extracting_geometry.m, extracting_shots.m, plot, stack_chart.m, mwigb.m and simage_display.m.

C H A P T E R  3

# Quality Control of Real Seismic Data

## 3.1  QUALITY CONTROL OF REFLECTION SEISMIC DATA

When seismic data is acquired and recorded, various quality control (QC) steps (called preprocessing in signal processing) are necessary to carry the remaining seismic data processing forward and ultimately obtain an accurate seismic image of subsurface structures. The QC process involves a series of steps to condition the data and prepare it for further quality control and processing. These steps include:

1. Demultiplexing: The data is transposed from the recording mode, where each record contains the same time sample from all receivers, to the trace mode where each record contains all time samples from one receiver. This is usually done in the field.

2. Reformatting: The data is converted from one seismic digital format (e.g., SEG-Y) to another format that is convenient to the processing software and used throughout the processing flow (e.g., MATLAB).

3. Setup of field geometry: The geometry of the field is written into the data (trace headers) in order to associate each trace with its respective shot, offset, channel, and CMP.

4. Trace editing: During this step, bad traces, or parts of traces, are muted (zeroed) or killed (deleted) from the data and polarity problems are fixed.

5. Gain application: Amplitude corrections are applied to account for amplitude losses due to spherical divergence and absorption.

6. Application of field statics: In land surveys, elevation statics are applied to bring the sources and receivers to a common datum level. This step can be delayed until the static correction process where better near-surface velocities might be available.

Steps 1-3 have already been taken care of for our data set while we are going to work on step 6 later on in Chapter 7. In the rest of this chapter, we will perform a simple trace editing step on the data set and then focus the discussion on gain application to our seismic data.

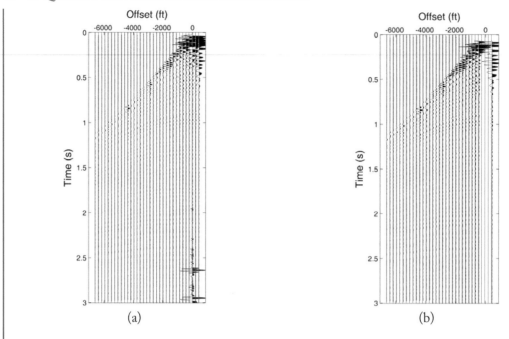

**Figure 3.1:** Seismic data shot gather number 16: (a) before and (b) after trace editing by muting.

## 3.2   TRACE EDITING

We are going to use the east Texas seismic data set where we already have noticed some high noisy amplitude traces from the previous chapter, particularly, in trace 31 of shot gather number 16. This is due to the use of a bad geophone, in this case, since we are dealing with land data. Such a recorded trace must be muted or, in other words, replaced by a zero trace. In MATLAB, there exist many ways to mute such bad traces but we simply can use the following step to do so on the complete data set (Figure 3.1):

```
1  [i,j]=find(D==max(max(D)));
2  D(:,j)=0;
```

## 3.3   AMPLITUDE LOSSES & THEIR CORRECTION

Among the various QC steps necessary to ultimately obtain an accurate seismic image of the subsurface structures is to correct for amplitude losses. Usually, seismic data in its raw state show a noticeable decrease in the amplitudes of its recorded traces with time (for example, see Figure 3.1(a)). It includes geometric divergence effects as waves spread out from a source as well as conversion of the seismic energy into heat besides other factors such as transmission losses [2]:

- Transmission loss: This occurs at each geological reflector where part of the propagating seismic incident waves will be reflected, refracted, diffracted, scattered, etc. There is no loss here in terms of the mechanical energy since the lost energy merely travels somewhere else.

- Geometric divergence: As the seismic wave spreads out from its source, its amplitude decays by an amount proportional to the reciprocal of the distance from the source to the location of the propagating seismic wave.

- Absorption: This occurs where the seismic energy is converted into heat by friction. This loss is proportional to the exponential of the distance from the source.

Now, amplitude correction or gain must be applied to seismic data sets at various stages. At the preprocessing stage, we may want to correct amplitudes due to geometric divergence as well as absorption losses. Also, whenever we want to display seismic data, we may want to boost weak signals by adding more gain to the data. There exist data independent and dependent amplitude correction schemes. The data-independent scheme corrects the amplitudes using a common scaling function to all the traces such as the gain through multiplication by a power of time using:

$$f_{corrected}(t) = f(t)t^{\alpha}, \tag{3.1}$$

where $f(t)$ is the seismic trace amplitude to be corrected, $t$ is the time independent variable, and $\alpha$ is the power of time which controls the change in the amplitude of $f(t)$. Another commonly used function is the exponential gain function correction:

$$f_{corrected}(t) = f(t)e^{\beta t}, \tag{3.2}$$

where $\beta$ is the parameter which controls the exponential function. The following MATLAB m-file uses the written function `iac.m` to apply independent amplitude correction using Equation 3.1 (the reader can also select to use the option T=1 for applying Equation 3.2) on shot gather 8:

```
1  shot_num=8;
2  p=0;
3  [Dshot,dt,dx,t,offset]=extracting_shots(D,H,shot_num,p);
4  pow=2;
5  T=0;
6  Dg=iac(Dshot,t,pow,T);
```

We can clearly see the amplitudes enhancements gained on this shot gather in Figure 3.2(b). We can further analyze the gain increase after the corrections by plotting the average trace amplitudes envelope in dB for the both shot gathers of Figure 3.2. Figure 3.2 shows such plots not only for the average trace (a) but also for trace number 33, both are generated using our written MATLAB function `seis_env_dB.m` as follows:

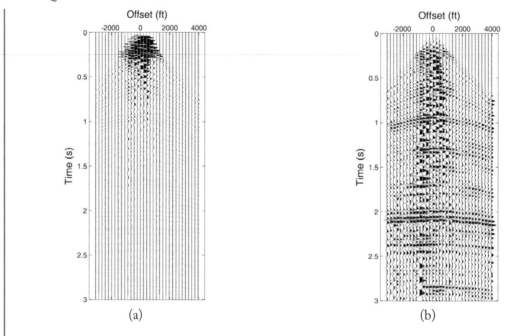

**Figure 3.2:** Seismic data shot gather number 8: (a) before and (b) after applying amplitude correction gain method of Equation 3.1.

```
1   tnum=33;
2   seis_env_dB(Dshot,Dg,t,tnum)
3   seis_env_dB(Dshot,Dg,t)
```

The data-dependent scheme, on the other hand, relies on multiplying each time sample by a scalar derived from a window of data around the sample. Such a technique is known as automatic gain control (AGC). Some of the famous AGC techniques include:

- RMS amplitude AGC: This method requires segmenting each trace into fixed time gates and then:

  1. Calculate the RMS value in each gate.

  2. Divide the desired RMS scaler by the RMS value of step 1 and multiply it by the amplitude of the sample at each gate center.

  3. Interpolate between these gate centers and multiply the result by the amplitude of samples corresponding in time.

- Instantaneous AGC: this is a bit different from the RMS AGC:

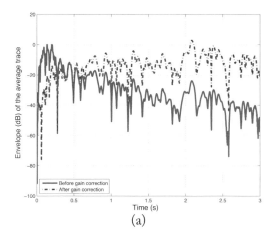
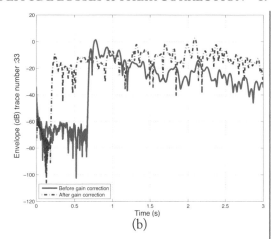

(a)                                (b)

**Figure 3.3:** The amplitude envelope gain in dB for (a) the average trace and (b) trace 33 both of Figure 3.2.

1. Calculate the absolute mean value for in a given gate of length $w$.

2. Divide the desired RMS scaler by the mean value of step 1 and multiply it by the amplitudes of all the samples in the gate.

3. Slide the gate down by one sample and repeat steps 1-2 until we have calculated all the amplitudes of all trace samples that have been corrected.

We use the MATLAB function `AGCgain.m` to apply AGC on shot gather 8 using the RMS AGC (Figure 3.4(a)) and the instantaneous AGC (Figure 3.4(b)). We generated these results via the following MATLAB scripts:

```
1  agc_gate=0.5;
2  T=1;
3  Dg=AGCgain(Dshot,dt,agc_gate,T);
```

```
1  agc_gate=0.5;
2  T=2;
3  Dg=AGCgain(Dshot,dt,agc_gate,T);
```

Note the amplitude envelope gain plots for the AGC'ed data in Figure 3.5. A question that may come into the readers' mind is: why do the AGC amplitude correction results show the triangular shapes of high amplitudes and amplitude gaps at the top of the figures? The answer is that these high-amplitude triangles are the result of noise that fills the whole AGC window (0.5 s in this case) before the arrival of seismic signals and dividing by their RMS value or absolute mean amplifies this low-amplitude noise. The amplitude gaps are due to the fact that the

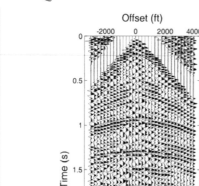
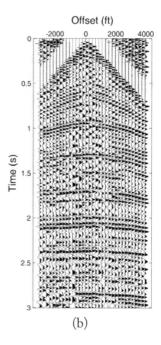

(a)   (b)

**Figure 3.4:** Seismic data shot gather number 8 after applying the AGC using the (a) RMS and (b) instantaneous based methods.

AGC window at these time shifts contains both low-amplitude noise and high-amplitude seismic signal and dividing by their RMS value or absolute mean amplifies the signal and attenuates the noise.

A final word of caution is that we must be careful when applying amplitude corrections techniques since they may destroy the signals character. At this preprocessing stage, Equation 3.1 with $\alpha = 2$, which is known as the $t^2$-correction, is used for geometrical spreading correction [6].

## 3.4   COMPUTER ASSIGNMENTS

1. Load the real seismic data file Book_Seismic_Data.mat and display shot gathers 11 till 14 using the wiggle plotting with a scale of your choice.

2. With $\alpha, \beta = 1.8, 2.2$ and 3.4, use both the multiplication by a power of time and the exponential gain function corrections on the selected shot gathers. Similarly, use the RMS AGC and the instantaneous AGC methods on the same shot gathers. Display and compare your results with the data before applying the required amplitude corrections. In your opinion, which method results in the best amplitude correction? Why?

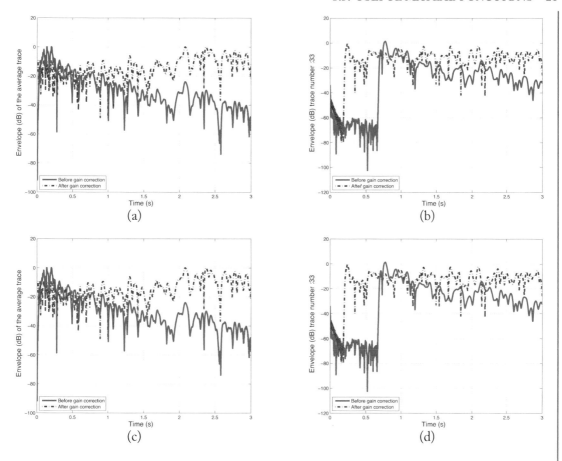

**Figure 3.5:** The amplitude envelope gain in dB for (a) the average trace and (b) trace 33 both of Figure 3.4(a). Similarly, the amplitude envelope gain in dB for (a) the average trace and (b) trace 33 both of Figure 3.4(b).

3. Mute the bad traces of shot gather 16 as in Section 3.2. Then apply the method of multiplication by a power of time with $\alpha = 2.0$ and all shot gathers and save the processed data with its header information as `Book_Seismic_Data_gain.mat` to be used later on.

## 3.5    USEFUL MATLAB FUNCTIONS

`seis_env_dB.m`, `iac.m` and `AGC_gain.m`.

CHAPTER 4

# Seismic Noise Attenuation

## 4.1 INTRODUCTION

Seismic data are highly corrupted with noise or unwanted energy arising from different kinds of sources. This unwanted energy can be classified into two main categories [2]: random noise (incoherent noise) and coherent noise. Seismic data processing, in general, can never eliminate all noise contaminating the data. Hence, the objective of seismic data processing is to improve as much as possible the signal-to-noise ratio (SNR). Using MATLAB examples, we will introduce some important means for analyzing the frequency (wavenumber) content of seismic data[1] and then attenuate some of the noise contaminating our seismic data using linear filtering.

## 4.2 SEISMIC SIGNAL & NOISE

Noise in seismic records is variable in both time and space [7]. Poor seismic records usually have SNR ratios less than one. One can define the signal of interest (coherent energy) as the energy which is coherent from trace to trace. Random noise, on the other hand, is the energy that is incoherent from trace to trace [8]. Furthermore, data from seismic events is correlated and its energy is generally concentrated in a fairly narrow band of, while noise is more uncorrelated and broadband [3]. However, this is only true for random noise. Spatially coherent noise is the most troublesome noise and can be highly correlated and sometimes aliased with the signal [8] and [9]. In general, noise can be considered as anything other than the desired signal. A more proper definition of noise contaminating seismic signals can be stated by defining the type of signals we are interested in. The authors in [8] define the signal of interest as the energy that is most coherent and desirable for geophysical interpretation of primarily reflected arrivals (signals). Anything other than that is considered to be unwanted energy, i.e., noise.

### 4.2.1 RANDOM NOISE

Disturbances in seismic data which lack phase coherency between adjacent traces are considered to be random noise. Unlike coherent noise energy, such energy is usually not related to the source that generates the seismic signals. In land seismic records, near-surface scatterers, wind, rain, and instrument are examples of sources generating random noise. Based on the assumption that random noise is an additive white Gaussian noise (AWGN) [2, 10], it can be attenuated easily in several different ways such as frequency filtering, deconvolution, wavelet denoising [11, 12, 13], filtering

[1]We are limiting the discussion here to $1 - D$ and $2 - D$. The same concepts can be used for the case of $3 - D$ seismic data.

using Gabor representation [14], stacking [1, 2] and many other methods. As discussed before in Chapter 1, stacking usually suppresses most of the incoherent noise and, therefore, improves the SNR by a factor of $\sqrt{M}$, where $M$ is equal to the number of stacked traces.

### 4.2.2    COHERENT NOISE

Spatially coherent noise is the energy which is generated by the source. It is an undesirable energy that is added to the primary signals. Such energy shows consistent phase from trace to trace. Examples of such a type in land seismic records are [2]: multiple reflections or multiples, surface waves like ground roll and, air waves, coherent scattered waves, dynamite ghosts, etc. Improper removal of coherent noise can affect nearly all the processing techniques and complicates interpretation of geological structures (see [2], [1], and [15]). There exist loads of techniques which deal with the problem of suppressing/attenuating coherent noise that contaminates seismic data [2, 15, 16, 17, 18, 19, 20, 21, 22]. Since our real seismic data contains mainly ground roll coherent noise, we are going to describe its main characteristics in the following subsection.

#### 4.2.2.1    Ground Roll Noise

They are surface (Rayleigh) waves traveling along the ground surface. They have generally low velocity ($< 1000$ m/s), low frequency ($< 15$ Hz), and high amplitudes. Their time-distance $t - x$ curves[2] are straight lines with low velocities and zero intercepts for an inline source. There might be several modes of ground rolls in the record because of their dispersive nature (i.e., different frequency components travel with different velocities). They are attenuated using source and receiver arrays in the field and various processing methods (e.g., frequency filtering, fan or $f - k$ filtering ). Figure 4.1 shows ground roll on shot gather number 8 from our real seismic data in the time-space $(t - x)$ (a), frequency-space $(f - x)$ (b), and frequency-wavenumber $(f - k)$ (c) domains. Clearly, we see the low frequency, low velocity and high amplitudes around the zero-offset location (indicated by the red lines in (a)).

## 4.3    SPECTRUM ANALYSIS AND FILTERING OF SEISMIC DATA

The filtering process is an important step in order to proceed further with the other seismic data processing steps (refer to typical seismic data processing workflow chart in Figure 1.3). The reader can examine references [2] and [4] for more information about seismic data filtering and noise suppression techniques. Since our land seismic data set contains ground roll noise, we simply can apply frequency linear filters such as band-pass filters (BPF)'s to attenuate their effect. In particular, BPF's enhance the overall gain of each seismic shot gather, and increases the SNR ratio by attenuating low and high frequency noise records, including the ground roll noise. However, before doing so, let us learn how to analyze the seismic data in the spectrum domain using

---

[2]A time-distance curve is a plot of wave arrival time against the source-to-geophone distance.

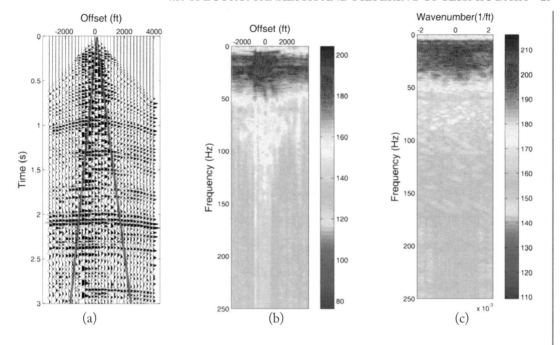

**Figure 4.1:** (a) Seismic data shot gather number 8 containing ground roll noise (red lines indicate ground roll noise). (b) Its $f - x$ and (c) its $f - k$ magnitude spectra. The color bars in (b) and (c) indicate the magnitude values in dB.

MATLAB. There exist different means for such useful analysis: the frequency (or wavenumber) content of one-dimensional $(1 - D)$ time (or space) signals, two-dimensional $(2 - D)$ like the frequency-space $(f - x)$ or the frequency-wavenumber $(f - k)$ spectra; each of which can be used to obtain meaningful interpretation and, therefore, apply a suitable filtering technique. These plots and their corresponding analysis and interpretations are very useful, particularly, when applying linear filtering to $1 - D$ and/or $2 - D$ seismic data sets.

Now, we can use the following MATLAB m-file containing the written M-functions `fx.m` and `fk.m` to, respectively, obtain the $f - x$ and $f - k$ magnitude spectra in dB as seen Figure 4.1(b) and (c):

```
1  clear,clc,close all
2  load Book_Seismic_Data_gain.mat
3
4  shot_num=8;
5  p=0;
6  [Dshot,dt,dx,t,offset]=extracting_shots(Dgz,H,shot_num,p);
```

```
7
8    scale=2;
9    mwigb(Dshot,scale,offset,t)
10   xlabel('Offset (ft)','FontSize',14)
11   ylabel('Time (s)','FontSize',14)
12
13   [Data_f,f]=fx(Dshot,dt);
14   figure,
15   pcolor(offset,f,20*log10(fftshift(abs(Data_f),1)))
16   shading interp;
17   axis ij;
18   colormap(jet)
19   colorbar
20   xlabel('Offset (ft)','FontSize',14)
21   ylabel('Frequency (Hz)','FontSize',14)
22   set(gca,'xaxislocation','top')
23   axis([min(offset),max(offset),0,max(f)])
24
25   [Data_fk,f,kx]=fk(Dshot,dt,dx);
26   figure,
27   pcolor(kx,f,20*log10(fftshift(abs(Data_fk))))
28   shading interp;
29   axis ij;
30   colormap(jet)
31   colorbar
32   xlabel('Wavenumber (1/ft)','FontSize',14)
33   ylabel('Frequency (Hz)','FontSize',14)
34   set(gca,'xaxislocation','top')
35   axis([min(kx),max(kx),0,max(f)])
```

Now, we design an Finite Impulse Response (FIR) BPF digital filter and apply it to each seismic trace in the shot gather number 8 using the written M-function bpf_fir.m as follows (with the given filter specifications):

```
1    N=100;
2    cut_off=[15,60];
3    [Dbpf,Dbpf_f]=bpf_fir(Dshot,dt,N,cut_off);
```

Figure 4.2 represents shot gather 8 before and after applying the above mentioned BPF on each trace along with the difference between both of them where clearly a great amount of the ground roll noise has been filtered. The $f - x$ magnitude spectra of Figure 4.2 is also seen in Figure 4.3 where the spectrum was banded between 15 and 60 Hz and the remaining was left out. A final word of caution is that although frequency filtering like BFP has improved the data SNR by attenuating the ground roll noise, it has also lowered the vertical resolution of the seismic data. This is common in seismic data processing and will be taken care of by applying deconvolution as we will explore in the next chapter.

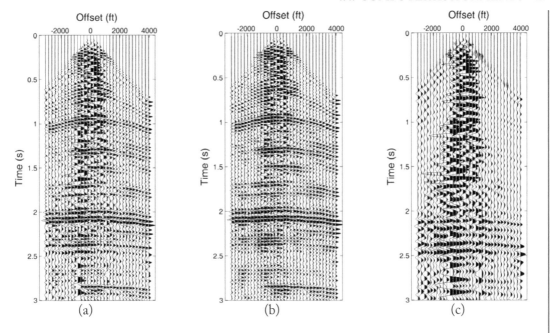

**Figure 4.2:** Seismic data shot gather number 8 containing ground roll noise (a) before and (b) after BPF filtering. (c) The difference between (a) and (b).

## 4.4    COMPUTER ASSIGNMENTS

1. Load the QC'ed real seismic data called `Book_Seismic_Data_gain.mat`. Extract and display shot gather 2 using the wiggle plotting with a scale of your choice. Using the MATLAB functions `fft` and `fftshift`, plot the magnitude and phase spectra for traces 1, 10, and 15 of this shot gather. Write down your own observations.

2. For the same shot gather, plot the magnitude and phase spectra of its $f - x$ and $f - k$ data representation. Use both linear and dB magnitudes. Comment on your results. In your opinion, what are the main differences in terms of interpretation between both the $f - x$ and the $f - k$?

3. Use the MATLAB's signal processing toolbox to design and apply the following filters:

   a. An FIR Low-pass filter (LPF) using any windowing method with a filter order of 50 and a cut-off of 15 Hz. Select any trace and apply your designed filter to it and display both the original trace and its LPF version. Comment on both traces by considering their time and frequency domain representations.

   b. Repeat (a) but with an Infinite Impulse Response (IIR) filter with an order of 5. Design the IIR filter using the bilinear transformation method.

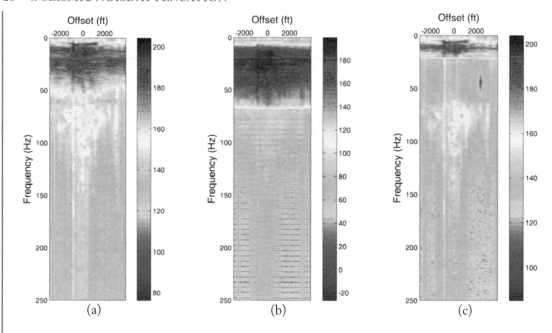

**Figure 4.3:** The $f - x$ magnitude spectra of the seismic data shot gather number 8 containing ground roll noise: (a) before and (b) after BPF filtering as well as for (c) the difference.

       c. Apply the FIR and IIR LPFs from parts (a) and (b), respectively, on each trace of shot gather 2. Display the original shot gather and the processed shot gathers. What can you notice from your results?

Note: here and in the next parts of this assignment, make sure that you use the same wiggle display scale of your filtered shot gathers in order to run fair comparisons.

4. Repeat Part 2 but with a high-pass filter (HPF) with a cut-off 60 Hz. In your opinion, which is better the LP filtered or HP filtered data results?

5. Apply band-pass filtering with cut-off values of 15 and 60 Hz using the M-function `bpf_fir.m` to all the shot gathers and save the resultant data set (with its header) as `Book_Seismic_Data_gain_bpf.mat` to be used in the coming chapters.

## 4.5    USEFUL MATLAB FUNCTIONS

`fx.m`, `fk.m`, `bpf_fir.m`, `abs`, `pcolor` and `log10`.

# CHAPTER 5

# Seismic Deconvolution

## 5.1 INTRODUCTION

After we have performed frequency filtering on seismic data via BPFs, the seismic data is smoothed and, hence, its vertical resolution is affected due to the loss of some of its original wider frequency band. One of the main aims of seismic deconvolution is to increase the vertical resolution of the data by compressing the source wavelet (to a zero-phase spike, if possible). This is known as *spiking deconvolution*. Seismic deconvolution, however, is not limited to this application where we can use it as well for noise attenuation such as multiples. Since our land seismic data set requires at this stage enhancement of its vertical resolution, we are going to elaborate here on spiking deconvolution and its application to our data set. However, before we get to deconvolution, we must define the seismic convolutional model.

## 5.2 THE SEISMIC CONVOLUTIONAL MODEL

The seismic convolutional model (recall Figure 1.1) is used to explain how the seismic trace is formed. The seismic convolutional model approximates the earth by a linear system. A linear system is one whose output $o(t)$ is given by the convolution of its input $i(t)$ with its response $r(t)$. That is, $o(t) = i(t) * r(t)$. In the seismic convolutional model:

- The system's output is $o(t) = s(t)$, where $s(t)$ is the recorded seismic trace.

- The system's input $i(t) = w(t)$, where $w(t)$ is the wavelet generated by the seismic source.

- The system's response $r(t) = e(t)$, where $e(t)$ is a series of impulses corresponding, in time and amplitude, to the reflection coefficients at layers boundaries. $e(t)$ is also known as the reflectivity (series). Therefore, according to this model, the seismic trace $s(t)$ is given by:

$$s(t) = w(t) * e(t). \tag{5.1}$$

A random noise component $\gamma(t)$, if present, is additive; hence, the noisy seismic trace becomes:

$$s_n(t) = w(t) * e(t) + \gamma(t). \tag{5.2}$$

The forward seismic convolutional model is used to compute synthetic seismograms $s(t)$ given the source wavelet $w(t)$ and earth's reflectivity $e(t)$ (Equations 5.1 and 5.2). The deconvolution (inverse seismic convolutional model) is used to:

1. Compute the earth's reflectivity $e(t)$ given the seismic trace $s(t)$ and the source wavelet $w(t)$. This is the most common objective of deconvolution. If the source wavelet is known, the deconvolution becomes deterministic. Whereas, if the source wavelet is not known, the deconvolution becomes statistical.

2. Compute the source wavelet $w(t)$ given the seismic trace $s(t)$ and the earth's reflectivity $e(t)$. This is used if a seismic trace is recorded near a borehole.

The seismic convolutional model is widely accepted because it agrees well with the observed seismic traces. The seismic convolutional model commonly assumes the following [2].

1. The earth is made up of horizontal layers of constant velocity.

2. The source generates only a primary wave, which is reflected on layer boundaries at normal incidence.

3. The source wavelet is stationary. That is, it does not change its shape as it travels in the subsurface.

4. The noise component $\gamma(t)$ is zero.

5. The Earth's reflectivity $e(t)$ is a white random series of impulses.

6. The seismic wavelet is a minimum-phase wavelet, which means that the wavelet has its energy concentrated at its start time.

Often, one or more of these assumptions might not be satisfied; in which case, advanced techniques of deconvolution have to be implemented.

## 5.3  SPIKING DECONVOLUTION AS A FILTERING PROCESS

We will illustrate this concept using the spiking deconvolution as an example. The aim of spiking deconvolution is to compress the source wavelet $w(t)$ into a zero-phase spike of zero width, i.e., $\delta(t)$. This means that we are eliminating the effect of the source wavelet and leave only the effect of the Earth's reflectivity in the seismogram. We can achieve this by convolving the seismic trace by the inverse filter, $h(t)$, of the source wavelet defined as:

$$w(t) * h(t) = \delta(t). \tag{5.3}$$

Taking the Fourier transform of Equation 5.3:

$$W(f)H(f) = 1, \tag{5.4}$$

where $W(f)$, $H(f)$, and $1$ are the Fourier transforms of $w(t)$, $h(t)$, and $\delta(t)$, respectively. From Equation 5.4, we can see that:

$$H(f) = 1/W(f) = [1/|W(f)|]\exp[-\phi_w(f)],$$

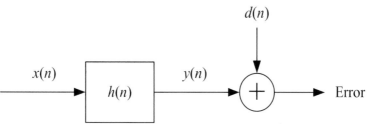

**Figure 5.1:** The Wiener optimum filter model.

which implies that $|H(f)| = 1/|W(f)|$, and $\phi_h(f) = -\phi_w(f)$. Note that $|W(f)|$ and $\phi_w(f)$ are the amplitude and phase spectra of $w(t)$ and $|H(f)|$ and $\phi_h(f)$ are the amplitude and phase spectra of $h(t)$. Therefore, the amplitude spectrum of the inverse filter is the reciprocal of that of the source wavelet whereas its phase spectrum is the negative of that of the wavelet. Taking the inverse Fourier transform of the $H(f)$, we get the desired inverse filter $h(t)$. Finally, spiking deconvolution is accomplished then by convolving the inverse filter $h(t)$ with the seismic trace $s(t)$:

$$h(t) * s(t) = h(t) * [w(t) * e(t)] = [h(t) * w(t)] * e(t) = \delta(t) * e(t) = e(t), \qquad (5.5)$$

which is the earth's response that we want to extract from the seismic trace.

In practice, the data is uniformly sampled and, therefore, we can assume that all the signals in Equation 5.2 are sampled and we can then rewrite it as:

$$s_n(n) = w(n) * e(n) + \gamma(n), \qquad (5.6)$$

where $n$ represents the time index. Hence, we can also use the $z$-transform in Equation 5.4, which results in $H(z) = 1/W(z)$. $H(z)$ is an infinite polynomial of $z$ that is convergent only if $w(n)$ is a minimum-phase wavelet. For practical reasons, the infinite polynomial $H(z)$ has to be truncated to $N$ terms (i.e., $F_N(z)$). Truncation generates less error if $w(n)$ is a minimum-phase wavelet and we include more terms of $H_N(z)$. The truncated filter $h_N(n)$ is calculated by taking the inverse $z$-transform of $H_N(z)$. Because of truncation, convolution of the truncated filter $h_N(n)$ with the wavelet will not give the desired output $d(n) = \delta(n) \equiv (1, 0, 0, \ldots)$. The actual output $y(n)$ is given by:

$$y(n) = h_N(n) * w(n). \qquad (5.7)$$

Apparently, $y(n) \neq d(n)$ and there will be a truncation error $E$ defined as:

$$E = \sum_n [d(n) - y(n)]^2. \qquad (5.8)$$

## 5.4   SPIKING DECONVOLUTION USING WIENER OPTIMUM FILTERS

Wiener optimum filtering involves designing a filter $h(n)$ so that the error $E$ in Equation 5.8 between the desired output $d(n)$ and the actual output $y(n)$ is minimum (refer to Figure 5.1). The actual output $y(n)$ is given as:

$$y(n) = h(n) * x(n), \tag{5.9}$$

where $x(n)$ is the input. Substituting equation 5.9 into 5.8:

$$E = \sum_n [d(n) - \{h(n) * x(n)\}]^2. \tag{5.10}$$

The goal is to compute the filter coefficients $(h(0), h(1), \ldots, h(N-1))$ so that the error $E$ is minimum, where the filter length, $N$, must be predefined. This is a typical least-squares problem, and the minimum error is attained by setting the partial derivative of $E$ with respect to $h(j)$ to zero:

$$\partial E / \partial h(j) = 0, \tag{5.11}$$

where $j = 0, 1, 2, \ldots, (N-1)$. Applying Equation 5.11 on 5.10 and simplifying, the result can be expressed as:

$$\sum_{n=0}^{N-1} r(j-n)h(n) = g(j) \tag{5.12}$$

for $j = 0, 1, 2, \ldots, (N-1)$. In matrix format, Equation 5.12 can be written as:

$$
\begin{bmatrix}
r(0) & r(1) & r(2) & \cdots & r(N-1) \\
r(1) & r(0) & r(1) & \cdots & r(N-2) \\
\vdots & \vdots & \vdots & \ddots & \vdots \\
r(N-1) & r(N-2) & r(N-3) & \cdots & r(0)
\end{bmatrix}
\begin{bmatrix}
h(0) \\
h(1) \\
\vdots \\
h(N-1)
\end{bmatrix}
=
\begin{bmatrix}
g(0) \\
g(1) \\
\vdots \\
g(N-1)
\end{bmatrix}
\tag{5.13}
$$

In Equations 5.12 and 5.13, $g(j)$ is the $j^{th}$ term of the crosscorrelation between $d(n)$ and $x(n)$:

$$g(j) = d(j) \otimes x(j), j = 0, 1, 2, \ldots, N-1. \tag{5.14}$$

Note that $g(n) = d(n) \otimes x(n) \neq x(n) \otimes d(n)$ and that we only need lags $0, 1, \ldots, (N-1)$. $r_{j-i}$ is the $j^{th}$ term of the autocorrelation of $x(n)$:

$$r(j) = x(j) \otimes x(j), j = 0, 1, 2, \ldots, N-1. \tag{5.15}$$

Note that $r(n) = x(n) \otimes x(n) = r(-n)$ and we only need lags $0, 1, \ldots, (N-1)$. If $g(j)$ and $r(j-\tau)$ are known, Equation 5.13 can be solved uniquely to find the filter $h(n)$. Equations 5.12 and 5.13 are known as the normal equations. The autocorrelation $(r(j))$ matrix of Equation 5.13 is a Toeplitz matrix that can be inverted efficiently using the Levinson recursion procedure.

As an application of the normal equations, we will use them to design the inverse filter of the source wavelet such that the desired output is a zero-lag spike . The input $x(n)$ is the source wavelet $w(n) = (w(0), w(1), \ldots, w(N-1))$ of length $N$, the desired output $d(n)$ is a zero-lag impulse function $\delta(n) = (1, 0, \ldots, 0)$ of length $2N - 1$, and the inverse filter is $h(n) = (h(0), h(1), \ldots, h(N-1))$ of length $N$. Therefore, the input autocorrelations are:

$$r(0) = w^2(0) + w^2(1) + \ldots + w^2(N-1)$$
$$r(1) = w(0)w(1) + w(1)w(2) + \cdots + w(N-2)w(N-1)$$
$$\cdots$$
$$r(N-1) = w(0)w(N-1).$$

The crosscorrelations on the right-hand side of the normal equations are:

$$g(0) = w(0)$$
$$g(1) = 0$$
$$\cdots$$
$$g(N-1) = 0.$$

Therefore, the normal equations become:

$$
\begin{bmatrix}
r(0) & r(1) & r(2) & \cdots & r(N-1) \\
r(1) & r(0) & r(1) & \cdots & r(N-2) \\
\vdots & \vdots & \vdots & \ddots & \vdots \\
r(N-1) & r(N-2) & r(N-3) & \cdots & r(0)
\end{bmatrix}
\begin{bmatrix}
h(0) \\
h(1) \\
\vdots \\
h(N-1)
\end{bmatrix}
=
\begin{bmatrix}
w(0) \\
0 \\
\vdots \\
0
\end{bmatrix}. \tag{5.16}
$$

The actual application of this theory to the exploration of subterranean layers, such as spiking and prediction deconvolution, began with the work of the Geophysical Analysis Group at the Massachusetts Institute of Technology (MIT) between 1960 and 1965, where it was a significant milestone in the history of seismic data processing [4, 23, 24].

## 5.5    THE TRACE-WAVELET RELATION

We just saw that we need to know the autocorrelation of the source wavelet in order to perform spiking deconvolution. However, for most impulsive sources, the source wavelet is not known and deconvolution or filtering cannot be performed directly. We need to either estimate the source wavelet or find relations between the amplitude spectra or autocorrelations of the seismic trace and source wavelet. Here, we investigate the latter case.

## 5.5.1    AMPLITUDE SPECTRA

The earth response $e(n)$ can safely be approximated as a white random series of impulses [2, 23]. If this is the case, the amplitude spectrum of $e(n)$ will be constant. That is:

$$|E(f)| \Longleftrightarrow E_0. \tag{5.17}$$

Using Equation 5.1 to find the trace amplitude spectrum and substituting Equation 5.17 yields:

$$|S(f)| = |W(f)||E(f)| = E_0|W(f)|. \tag{5.18}$$

Equation 5.17 means that the amplitude spectrum of the seismic trace is a scaled version of the amplitude spectrum of the source wavelet.

## 5.5.2    AUTOCORRELATIONS

Because of the random nature of $e(n)$, its autocorrelation $r_e(n)$ is generally zero everywhere except at $n = 0$, where it is equal to the energy in $e(n)$ given by:

$$r_{e0} = \sum_n e^2(n) = e^2(0) + e^2(1) + \dots \tag{5.19}$$

Therefore, $r_e(n)$ can be approximated as:

$$r_e(n) = r_{e0}\delta(n). \tag{5.20}$$

Using this fact about $r_e(n)$, we can derive a relation between the trace and wavelet autocorrelations ($r_s(n)$ and $r_w(n)$) as follows:

$$
\begin{aligned}
r_s(n) &= s(n) \otimes s(n) \\
&= s(n) * s(-n) \\
&= [w(n) * e(n)] * [w(-n) * e(-n)] \\
&= [w(n) * w(-n)] * [e(n) * e(-n)] \\
&= r_w(n) * r_e(n) \\
&= r_{e0}(n) * r_w(n) \\
&= r_{e0}r_w(n). \tag{5.21}
\end{aligned}
$$

Equation 5.21 means that the autocorrelation of the seismic trace is a scaled version of the autocorrelation of the source wavelet. The benefit of Equations 5.19 and 5.21 is that we can use $|S(f)|$ or $r_s(n)$ whenever $|W(f)|$ or $r_w(n)$ are needed in deconvolution or filtering. Provided the randomness and whiteness assumptions about $e(n)$ are satisfied, all we miss by using $|S(f)|$ or $r_s(n)$ instead of $|W(f)|$ or $r_w(n)$ is a scaling factor.

## 5.6    SPIKING DECONVOLUTION IN PRACTICE

To perform spiking deconvolution in practice, we need to set the following parameters:

- Autocorrelation window ($w$): This sets up the part of seismic trace from which we will select the elements of the autocorrelation matrix in the normal equations.

- Filter length ($N$): This sets up the length of the spiking filter $h(n)$.

- Percent prewhitening ($\varepsilon$): This sets up the amount of white random noise we want to include into our autocorrelation matrix to stabilize the solution of the normal equations.

## 5.6.1    AUTOCORRELATION WINDOW

The choice of deconvolution parameters depends largely on the characteristics of the autocorrelation of the seismic trace. Therefore, it is important to choose a suitable autocorrelation window (gate) that will be used to calculate the deconvolution parameters. The autocorrelation window should include the part of the record that contains useful reflection signal and should exclude coherent (e. g., ground roll) or incoherent noise (e.g., later parts of the record). The length of the autocorrelation window should be greater than eight times the largest operator length that will be used for that data set [2]. Here, we use the whole time window length of the data for spiking deconvolution.

## 5.6.2    FILTER LENGTH

The filter (operator) length should be equal to the wavelet length. The first transient zone of the trace autocorrelation is the part that mostly represents the autocorrelation of the source wavelet. The first transient zone is the first part of the autocorrelation that contains high amplitudes. The operator length should be selected so that it is approximately equal to the length of the first transient zone of the trace autocorrelation. The optimum operator length should not leave considerable amount of energy in the trace autocorrelogram.

## 5.6.3    PERCENT PREWHITENING

The amplitude spectrum of the spiking deconvolution operator (inverse filter) is the reciprocal of that of the source wavelet. If the amplitude spectrum of the source wavelet is zero at some frequencies, then the amplitude spectrum of inverse filter will be unstable (i. e., it will be infinite at these frequencies). A similar numerical instability might be encountered when inverting the trace autocorrelation matrix if the determinant of the autocorrelation matrix is zero. To ensure numerical stability, we introduce an artificial level of white random noise into the trace amplitude spectrum and autocorrelation before deconvolution. This process is called prewhitening.

Prewhitening is achieved by adding a white random noise, with a very small variance ($\varepsilon$), to the trace amplitudes at every time sample. This is equivalent to adding a very small positive constant to the zero-lag autocorrelation ($r(0)$ in Equation 5.17) of the trace [2]. It is also equivalent to adding a very small positive constant to the amplitude spectrum of the trace at every frequency component. The magnitude of prewhitening is measured as a percentage of the zero-lag autocorrelation value

$r(0)$. In practice, typically 0.1% to 0.3% (i.e., $0.001r(0)$ to $0.003r(0)$) prewhitening is standard in processing.

### 5.6.4   SPIKING DECONVOLUTION OF OUR DATA SET

We use here our written M-functions `spiking_decon.m` and `auto_correlation_map.m` to perform spiking deconvolution filtering based on 5.17 and calculate the autocorrelogram, respectively. The m-file is as follows:

```
1   load Book_Seismic_Data_gain_bpf.mat
2
3   shot_num=4:6;
4   p=1;
5   [Dshot,dt,dx,t,offset]=extracting_shots(Dbpf,H,shot_num,p);
6   [nt,nx]=size(Dshot);
7
8   scale=5;
9   mwigb(Dshot,scale,offset,t)
10  xlabel('Trace number','FontSize',14)
11  ylabel('Time(s)','FontSize',14)
12
13  max_lag=0.2;
14  [Dauto,lags]=auto_correlation_map(Dshot,max_lag,dt);
15  scale=5;
16  mwigb(Dauto,scale,offset,lags)
17  xlabel('Trace number','FontSize',14)
18  ylabel('Time lag (s)','FontSize',14)
19
20  mu=0.1;
21  Ds=spiking_decon(Dshot,max_lag,mu,dt);
22  scale=5;
23  mwigb(Dt,scale,offset,t)
24  xlabel('Trace number','FontSize',14)
25  ylabel('Time (s)','FontSize',14)
```

Figure 5.2 shows the first three shot gathers of the data before applying spiking deconvolution. Their autocorrelograms are shown in Figure 5.3, while Figure 5.4 shows them after applying spiking deconvolution. We notice that the data became more spiky and we can further analyze this in the spectrum domain via the power spectral density of the average traces as we see in Figure 5.5. This figure was generated using the following commands:

```
1   Davg_before=mean(Dshot');
2   fs=1/dt;
3   [Davg_before,f] = periodogram(Davg_before,[],2*nt,fs);
4   Davg_before=Davg_before/max(Davg_before);
5   Davg_before=20*log10(abs(Davg_before));
6
7   Davg_after=mean(Ds');
8   [Davg_after,f] = periodogram(Davg_after,[],2*nt,fs);
```

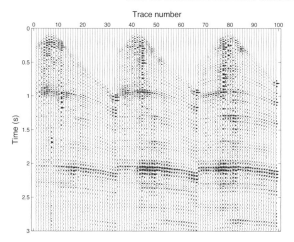

**Figure 5.2:** Shot gathers: 4, 5 and 6 before applying spiking deconvolution.

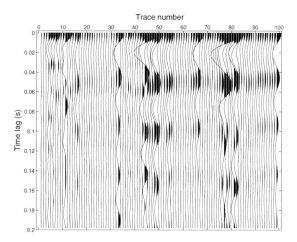

**Figure 5.3:** Autocorrelograms of shot gathers: 4, 5 and 6.

```
 9  Davg_after=Davg_after/max(Davg_after);
10  Davg_after=20*log10(abs(Davg_after));
11
12  f=linspace(-0.5,0.5,2*nt)/dt;
13  figure,plot(f,Davg_before,f,Davg_after,'r--')
14  xlabel('Frequency (Hz)','FontSize',14)
15  ylabel('Normalized PSD','FontSize',14)
16  grid
17  legend('Before decon','After decon')
```

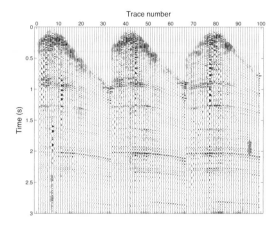

**Figure 5.4:** Shot gathers: 4, 5 and 6 after applying spiking deconvolution.

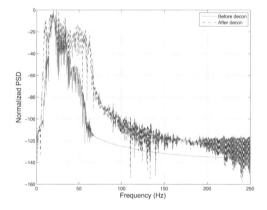

**Figure 5.5:** PSD of the average trace of shot gathers: 4, 5 and 6 before and after spiking deconvolution.

Finally, to further enhance the results before we sort the data in the next chapter, we have applied instantaneous AGC with window length of 0.5 s to the deconvolved data as seen in Figure 5.6. We applied AGC to compensate for the lost amplitudes after deconvolution.

## 5.7    COMPUTER ASSIGNMENTS

1. Apply spiking deconvolution using the M-function `spiking_decon.m` followed by an instantaneous AGC to all the shot gathers and save the resultant data set (with its header) as `Book_Seismic_Data_gain_bpf_sdecon_gain.mat` to be used in the coming chapters.

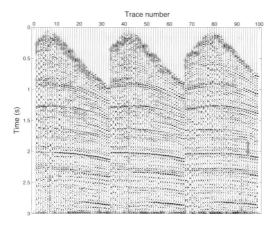

Figure 5.6: Shot gathers: 4, 5 and 6 after applying spiking deconvolution and instantaneous AGC.

2. Use the formulation of Equation 5.13 to generalize the M-function `spiking_decon.m` so it can be used for predication deconvolution. For information, we recommend the reader to look into [2].

## 5.8    USEFUL MATLAB CODES

`spiking_decon.m`, `auto_correlation_map.m`, and `periodogram`.

CHAPTER 6

# Carrying the Processing Forward

## 6.1 INTRODUCTION

After we have performed various seismic data processing steps such as: data quality control, frequency filtering and deconvolution on the east Texas real seismic data, we are ready now to prepare the data for revealing a true image of the subsurface. This requires compressing the seismic data to infer a first approximation for such image and may include:

- sorting the shot gathered data into common mid-point gathers (CMP);

- picking appropriate stacking velocities and applying accordingly normal move-out corrections (NMO); and

- stacking all CMPs, i.e., average each CMP traces to form one stacked CMP trace and then concatenate all stacked CMP traces together. This will be our first approximation for the subsurface image of interest called the stacked section.

Therefore, we are going in this chapter to briefly explain and apply to our data each process alone in order to obtain a stacked section.

## 6.2 COMMON MIDPOINT SORTING

Surface seismic reflection surveys are commonly acquired using the common midpoint (CMP) method. In this method, points in the subsurface are covered more than once by primary reflections from different source-receiver pairs. See Figures 6.1 and 6.2 for examples of shot gather and CMP configurations. Other names for this method include common reflection point (CRP) and common depth point (CDP). Traces reflected from the same midpoint form a CMP gather while the number of traces in a gather is called the fold of that gather. Common CMP folds are 80, 120, 240, and 480. We note here that the CMP spacing is half of the trace (receiver) spacing in a survey.

Seismic data is acquired in the shot gather mode while most seismic data processing is performed in the CMP-offset mode. Therefore, we need to sort the traces between these modes. Generally, we need a way to sort the traces into other modes as well. For this purpose, we use stacking charts. A stacking chart is a chart in which the x-axis indicates the geophone location and the y-axis indicates the source location. It is used to sort the traces into various modes or gathers such as shot, receiver, offset, or CMP. On a stacking chart (refer to Figure 6.3 which shows the data stacking chart):

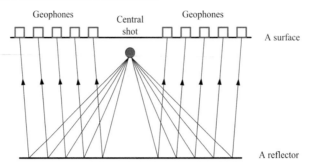

**Figure 6.1:** Two-sided common shot gather configuration where five receivers from each side of the shot record.

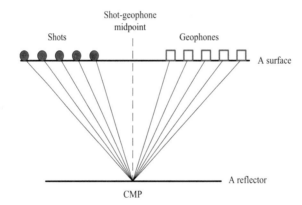

**Figure 6.2:** Common mid-point gather (CMP) configuration. We see that each of the shot-receiver combinations is centered around the same mid-point.

- Points along one diagonal have a common midpoint (the gather is called a common midpoint (CMP) gather).

- Points along the other diagonal have a common offset (the gather is called a common offset gather (COG)).

- Points along a horizontal line have a common source (the gather is called a common source gather).

- Points along a vertical line have a common receiver (the gather is called a common receiver gather (CRG)).

In order to sort the our seismic data set, we can use the M-function ssort.m as follows:

```
1  load Book_Seismic_Data_gain_bpf_sdecon_gain.mat
```

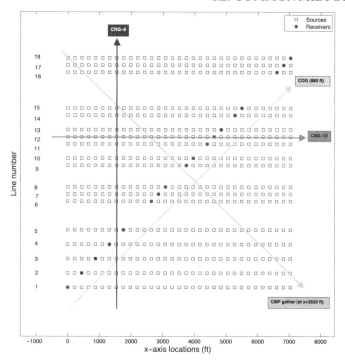

**Figure 6.3:** The stacking chart plot of the seismic data.

```
2   [Dsort,Hsort] = ssort(Ds_gain,H);
3   save Book_Seismic_Data_gain_bpf_sdecon_gain_sorted.mat Dsort Hsort
```

Note that we have saved the sorted file so it will be easily accessible later on. Furthermore, if we run the following script:

```
1   [cmps,fold_cmp]=extracting_cmp_fold_num(Dsort,Hsort);
2   figure,stem(cmps,fold_cmp,'-')
3   xlabel('CMP numbers','FontSize',14)
4   ylabel('Fold','FontSize',14)
5   set(gca,'YMinorGrid','on')
```

we obtain the fold or number of traces per CMP as shown in Figure 6.4 based on the M-function `extracting_cmp_fold_num.m`. From this figure, we see that the data starts at CMP 203 with a single trace then it increases until it reaches the maximum fold of 18 at CMP number 235 and then ends at CMP 266 with one trace. This is also confirmed by the stacking chart in Figure 6.3. Figure 6.5 shows several CMP gathers extracted from the sorted data and its header displayed using the following script:

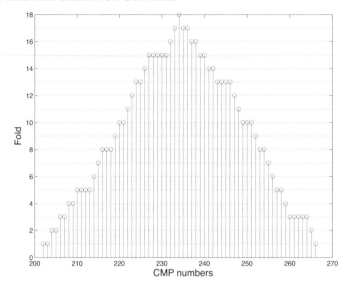

**Figure 6.4:** The fold (number of traces per CMP) versus the CMP numbers.

```
1  cmp_num=208;
2  [Dcmp,t,cdp]=extracting_cmp(Dsort,Hsort,cmp_num);
3  scale=2;
4  mwigb(Dcmp,scale,cdp,t)
5  xlabel(['CMP: ',num2str(cmp_num),''],'FontSize',14)
6  ylabel('Time(s)','FontSize',14)
```

Note the number of traces per CMP gather varies from a CMP gather to another and is consistent with Figure 6.4.

## 6.3    VELOCITY ANALYSIS

The objective of velocity analysis is to determine the seismic velocities of layers in the subsurface. Seismic velocities are used in many processing and interpretation stages such as: spherical divergence correction, NMO correction and stacking, interval velocity determination, migration, and time to depth conversion. There are different types of seismic velocities such as: the NMO, stacking, RMS, average, interval (Dix), phase, group, and migration velocities. The velocities that can be derived reliably from the time-space $(t - x)$ data (such as those in Figure 6.5) are the NMO, RMS, and stacking velocities. We are going to work with stacking velocities which will be used afterwards for NMO correction of our data.

We can obtain or, to be more precise, pick stacking velocities by running a velocity analysis on a given CMP gather, where we aim to obtain velocities that flatten out hyperbolas (which mainly

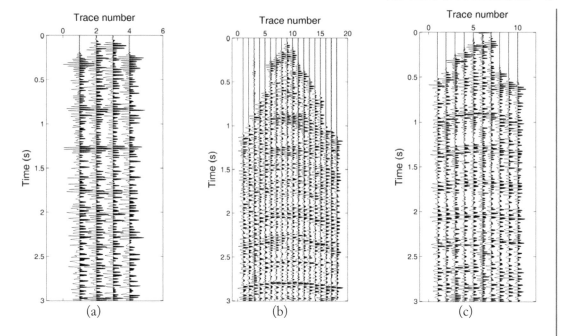

**Figure 6.5:** CMP gathers number (a) 208, (b) 234, and (c) 250 of our CMP-sorted seismic data.

represent reflections) on CMP gathers. This will increase the energy contained in a seismic event after adding all the traces of that particular CMP together, i.e., we increase the power of the overall stack.

There exist many means that we can use to compute the stacking velocities such as the *the velocity spectrum and the constant velocity stack* techniques. Both are commonly used and we will select the velocity spectrum technique to obtain the stacking velocities of our data.

### 6.3.0.1  The Velocity Spectrum

This method attempts to find the stacking velocity to each reflector. It maps the time-space data of a single CMP gather onto a velocity-spectrum plane. In the velocity-spectrum plane, the vertical axis is $t_0$ and the horizontal axis is $v_s$. The method consists of the following steps.

1. Select a CMP gather that has a relatively high SNR ratio. The CMP gather should be sorted in offset.

2. Determine the minimum (usually equals 0) and maximum (usually equals to the record length) $t_0$ that you want to analyze.

3. Determine the minimum, maximum, and increment of $v_s$ to be attempted.

4. Determine the gate width, $w$, around the reference time $t_0$. This is usually equal to the dominant period of the data.

5. Start with the minimum $t_0$ and $v_s$.

6. Compute $t(x) = \sqrt{t_0^2 + x^2/v_s^2}$, where $t_0$ and $v_s$ are set to the minimum $t_0$ and minimum $v_s$ of step (5) and $x$ is the offset of the traces in that CMP gather.

7. The amplitudes in a gate of width $w$ centered about $t(x)$ calculated from step (6) are selected from all the traces in the gather. We will call the amplitude of a time sample within this gate $A_{ij}$, where $1 \le i \le N$, $1 \le j \le M$, $N$ is the number of samples in the gate $w$, and $M$ is the number of traces in the CMP gather.

8. The sum of the amplitudes corresponding to the first time sample of the gate on all traces in the gather is computed and squared $(\sum_{j=1}^{M} A_{ij})^2$.

9. Step 8 is repeated for all the time samples in the gate $w$.

10. The squared sums are added together to give the stack energy $E_s = \sum_{i=1}^{N}(\sum_{j=1}^{M} A_{ij})^2$.

11. Now, sum up the squared amplitudes of the first sample on every trace .

12. Repeat step (11) for all the other samples in the gate.

13. The sums of the squares are added together to give the prestack energy $E_u = \sum_{i=1}^{N} \sum_{j=1}^{M} A_{ij}^2$.

14. Calculate the semblance $NE = (1/M)(Es/Eu)$. Note that: $0 \le NE \le 1$, and that it is larger if the amplitudes in the gate are aligned following a hyperbola whose $t_0$ and $v_s$ are equal to those of the hyperbola you are currently fitting.

15. Now you have one point on the velocity-spectrum plane, namely (minimum $t_0$, minimum $v_s$, $NE$).

16. While fixing $t_0$, increment $Vs$ and repeat steps (6)-(14) until you reach the maximum $v_s$.

17. Increment $t_0$ by $L = N/2$ samples and repeat steps (5)-(16) until you reach the maximum $t_0$.

18. For a reflection that has a zero-offset two-way traveltime[1] equal to $t_0$, its correct $v_s$ is the one that is associated with the maximum semblance occurring at that $t_0$.

19. Select another CMP gather and repeat steps (5)-(17).

---

[1]Two-way travel time is the total time it takes a seismic wave to travel down from the source to a reflection point and back up to the receiver.

Applying these steps should end up with a set of picks $(t_0, v_s)$ for every selected CMP. We do not pick for each CMP. Hence, to find the $(t_0, v_s)$ sets for the other, unprocessed CMPs, we interpolate them.

The parameters $Es$, $Eu$, and $NE$ are measures of coherency (similarity) of the signal along a hyperbolic curve [25]. The coherency measure is usually displayed as a contour plot. Other important parameters to consider when using the velocity spectrum method are the minimum, maximum, and increment $v_s$. The velocity spectrum method is more suited for noise-contaminated data sets.

The following script uses the M-function vel_picking.m that will interact with its user through the semblance plot and the computer mouse. The user will pick possible velocities by pointing and clicking the pointer on the semblance plot. It will also display the selected CMP gathers along with their semblance as shown, for example, in Figure 6.6(a):

```
1  load Book_Seismic_Data_gain_bpf_sdecon_gain_sorted.mat
2
3  cmp_step=5;
4  cmp_start=205;
5  cmp_end=255;
6  vmin = 5000;
7  dv=200;
8  nv = 51;
9  n_pts=8;
10
11 [v_stack,t_stack]=vel_picking(Dsort,Hsort,vmin,dv,nv,cmp_start,...
12 cmp_end,cmp_step,n_pts);
13
14 save Book_Seismic_Data_gain_bpf_sdecon_gain_sorted_velocities.mat
15 v_stack t_stack cmp_step cmp_start cmp_end
```

The output of the M-function vel_picking.m is composed of two vectors: one containing the stacking times and the other containing the stacking velocities, where we are going to use them for NMO correction in the coming section.

## 6.4    NORMAL MOVEOUT (NMO) CORRECTION

NMO correction will apply the picked stacking velocities to CMP traces and corrects with the non-zero offset traveltimes for their additional traveltime from source to receiver. The goals of NMO correction are:

- Preparation of the data for stacking.

- Estimating the NMO velocity function.

The normal moveout $\Delta t_{NMO}(x)$ is defined as the time difference (on the seismic section) between the two-way traveltime, $t(x)$, at an offset $x \neq 0$ and the two-way traveltime at zero-offset (i.e., $x = 0$) $t_0$:

$$\Delta t_{NMO}(x) = t(x) - t_0. \tag{6.1}$$

In the case of a single horizontal layer with constant velocity, the time-distance $(t - x)$ curve is exactly a hyperbola given by:

$$t^2(x) = t_0^2 + x^2/v^2, \tag{6.2}$$

where $v$ is the layer velocity. We can show that the NMO-correction, in this case, is given by:

$$t_{NMO}(x) \approx \frac{x^2}{2t_0 v^2}, \tag{6.3}$$

where the approximation is better for small offsets or where the maximum offset over the reflector depth is less than 2. For NMO correction, $\Delta t_{NMO}(x)$ is subtracted from $t(x)$ such that the two-way traveltime at offset $x$ after NMO correction, $t_{NMO}(x)$, is approximately equal to $t_0$:

$$t_{NMO}(x) = t(x) - \Delta t_{NMO}(x) \approx t_0. \tag{6.4}$$

Equation 6.3 implies that $\Delta t_{NMO}(x)$ increases with offset and decreases with depth (or $t_0$) and velocity. We can see from Equation 6.4 the following.

- If the correct NMO velocity is used for NMO correction, the event will be horizontally aligned at $t = t_0$.

- If a higher velocity is used for NMO correction, then the event will be undercorrected (i.e., concave down).

- If a lower velocity is used for NMO correction, then the event will be overcorrected (i.e., concave up).

For multiple horizontal constant-velocity layers, the $t - x$ curve is not a hyperbola, but we approximate it by a hyperbola. This approximation is better for small offsets. Therefore, we replace the multiple layers with one layer with an average velocity that will produce the closest $t - x$ curve to a hyperbola. This average velocity is called the stacking or NMO velocity. For a dipping layer, or a set of dipping layers, the NMO velocity to any dipping reflector increases with its dip angle. However, for small dip angles ($< 15$), the effect of dip on the NMO velocity is negligible. Therefore, for multiple horizontal and gently dipping layers (as the case of our data), we use equation 6.4 for NMO correction. For highly dipping layers, we must correct for the effect of dip on velocity. We do this by first Dip Moveout (DMO) processing then we perform the NMO correction. We recommend the reader to look into [2] for more DMO details.

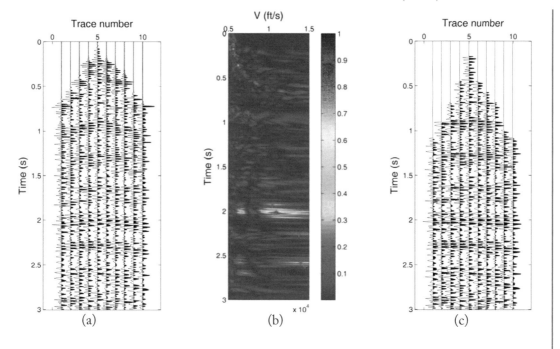

**Figure 6.6:** (a) CMP gather 220 of our sorted seismic before NMO correction. (b) Its semblance and (c) after correction.

### 6.4.0.2 NMO Stretching

The NMO correction maps the data in a CMP gather from the $x - t(x)$ plane to the $x - t_{NMO}(x)$ plane using equation 6.4. However, it does not shift the traces up in time. Because both the velocity and $t_0$ increase downward, the NMO correction applied (using equation 6.3) for a later $t_0$ will be smaller than that applied at an earlier $t_0$. Therefore, two points on the same event with a time separation equal to $\Delta t_b$ before NMO correction will have a separation of $\Delta t_a > \Delta t_b$ after NMO correction because the upper point will experience larger NMO correction than the lower one. We call this NMO stretching and it causes a frequency distortion to lower frequencies. Stretching is quantified as:

$$S_{NMO}(x) = \frac{\Delta t_{NMO}(x)}{t_0} \approx \frac{1}{2}(\frac{x}{t_0 v})^2. \tag{6.5}$$

NMO stretching is mainly confined to large offsets and shallow times as we can see from equation 6.5. Stacking NMO-corrected and stretched traces will severely damage the shallow seismic events. Therefore, we have to remove the stretching before we stack. We do this by muting the considerably stretched zones from the gather. The muted zone is usually set to a threshold in terms of $S_{NMO}$ so that zones with $S_{NMO}$ greater than the threshold will be muted. Typical values for mute threshold are between 0.5–1.50. Selecting a small threshold value avoids stretching but might cause excessive

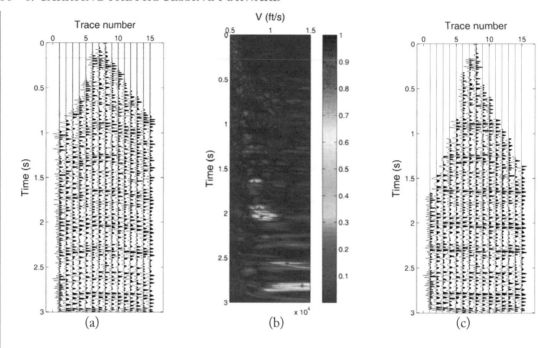

**Figure 6.7:** (a) CMP gather 230 of our sorted seismic before NMO correction. (b) Its semblance and (c) after correction.

loss of data, while selecting a large threshold value avoids loss of data but might introduce excessive stretching of the data.

The following MATLAB script uses the M-function called `nmo_correction.m` to perform NMO correction on seismic data CMP gathers based on the estimated stacking velocities from the previous section:

```
1  load Book_Seismic_Data_gain_bpf_sdecon_gain_sorted.mat
2  load Book_Seismic_Data_gain_bpf_sdecon_gain_sorted_velocities.mat
3  max_stretch=10;
4  [Dsort,Hsort]=nmo_correction(Dsort,Hsort,v_stack,t_stack,...
5  cmp_start,cmp_end,cmp_step,max_stretch);
6  save Book_Seismic_Data_gain_bpf_sdecon_gain_sorted_nmo_corrected
7  Dsort Hsort
```

Figures 6.6(c), 6.7(c), and 6.8(c) show various NMO-corrected CMP gathers. We can notice that the hyperbolic seismic events have become more flat for different layers. Now, we are ready to obtain our first approximation of the subsurface image using stacking. Note that the maximum stretch value

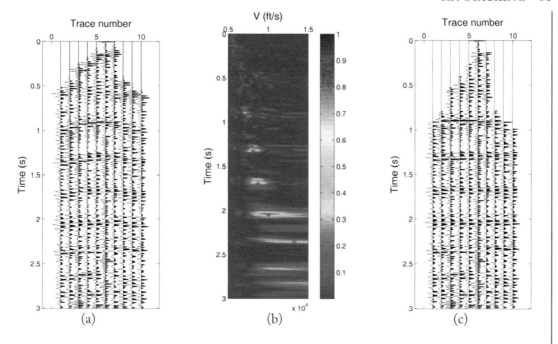

**Figure 6.8:** (a) CMP gather 250 of our sorted seismic before NMO correction. (b) Its semblance and (c) after correction.

we used in our NMO M-function is equal to 50. A very good exercise for the user is to vary the maximum stretch value and see how it will affect the results.

## 6.5    STACKING

The purpose of stacking is to enhance the SNR ratio by eliminating coherent and incoherent noise in the data and to reveal a first subsurface image approximation. The traces in the NMO-corrected CMP gather are stacked (summed up) to produce one stacked trace that represents that CMP. The amplitude of the stacked trace can be the sum or average of the amplitudes of traces in the CMP gather. As we mentioned before, stacking $M$ traces enhances the SNR ratio by $\sqrt{M}$. The stacked section is displayed with the CMP number along the horizontal direction, and $t_0$ along the vertical direction as we see in Figure 6.9 based on the following script which stack the data using the M-function `sstack.m`:

```
1  load Book_Seismic_Data_gain_bpf_sdecon_gain_sorted_nmo_corrected
2  [Dstacked,t,cmp_num]=sstack(Dsort,Hsort);
3  save
4  Book_Seismic_Data_gain_bpf_sdecon_gain_sorted_nmo_corrected_stacked
```

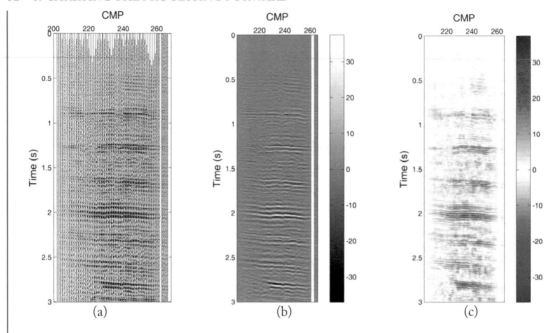

**Figure 6.9:** The stacked section of our seismic data in various displays: (a) variable area-wiggle, (b) variable density in gray and (c) variable density in colors. Recall that the color bars in (b) and (c) refers to the amplitude dynamic range of the data.

```
5  cmp_num t Dstacked
```

Clearly, we can see the continuity of "geological" layers but yet this stacked section requires us to improve its horizontal (spatial) resolution. We are going to perform one more seismic data processing step, namely, migration, to do that in the next chapter.

## 6.6   COMPUTER ASSIGNMENTS

1. Display using the wiggle variable area mean the 18 shot gathers and the 65 CMP gathers, one at a time and analyze this carefully with Figure 6.3. You can use the MATLAB capabilities to make a movie for each.

2. Select CMP gather number 250 along with its stacking time and velocities and then:

   • Add 1500 ft/s to the velocity values and perform NMO correction.
   • Subtract 1500 ft/s to the velocity values and perform NMO correction.

   What can you realize from both NMO corrected CMP gathers? Explain.

# 6.7 USEFUL MATLAB CODES

`ssort.m`, `extracting_cmp.m`, `extracting_cmp_fold_num.m`, `vel_picking.m`, `nmo_correction.m`, and `sstack.m`.

# CHAPTER 7

# Static Corrections

Static corrections are applied to seismic data in order to compensate for various effects on the data such as those related to near surface, variations in elevations, weathering and reference to a datum [2]. By applying static corrections, we aim to determine the reflection time assuming the existence of a flat surface plane with no weathering layer. These corrections include:

- Elevation static correction, which accounts for variable elevations of the sources and receivers.

- Residual static correction, which accounts for lateral variations in the velocity and thickness of the weathering layer (WL).

The WL is the shallowest low-velocity layer. It is composed of unconsolidated and loosely consolidated sediments. Common WLs in arid areas include: sand dunes, sabkhas, gravel plains, karsts, and valley fills. The WL velocity is usually much less than those of the underlying sub-WL (bedrock) and deeper layers. Therefore, the WL produces a large contribution to the overall traveltime of rays.

## 7.1 ELEVATION STATIC CORRECTION

Elevation static correction involves the computation and removal of the effect of different source and receiver elevations. It is also known as field, datum, and topographic correction. It involves bringing the sources and receivers to a common datum, usually below the elevation of the lowest source or receiver. For this, we need a replacement velocity ($V_r$) for the material between the datum and the source or receiver. The replacement velocity is either assumed from prior knowledge of the area or can be estimated from uphole times or direct arrivals. The elevation static correction ($T_D$) is given by [2]:

$$T_D = \frac{(E_S - Z_S - E_D) + (E_R - Z_R - E_D)}{V_r}, \tag{7.1}$$

where $E_S$ is the ground elevation at shot location (from mean sea level), $Z_S$ is the depth of shot (and is equal to 0 for a surface source), $E_R$ is the ground elevation at receiver location (from mean sea level), $Z_R$ is the depth of receiver (and is equal to 0 for a surface geophone), and $E_D$ is the datum elevation (from mean sea level). $T_D$ is always subtracted from the two-way traveltime of the trace belonging to that particular source-receiver pair.

## 7.2 RESIDUAL STATIC CORRECTION

After the elevation static correction, it is important to correct for the effect of variable thickness and lateral velocity variation of the WL. The main methods used to correct for these effects are [2]:

1. uphole surveys

2. refraction statics

3. surface-consistent statics

## 7.2.1   UPHOLE SURVEYS

In this method, a fairly shallow hole (100-200 m) that penetrates the WL and the upper part of the sub-WL is drilled for this purpose. Several geophones are placed at various (known) depths in the hole. The geophone locations must span the weathering and sub-WLs. A shot is fired at the surface near the hole and the direct traveltimes to the geophones are recorded. A plot of the direct traveltimes versus the geophone depths can be used to compute the velocities of the weathering and sub-WLs as well as the thickness of the WL at the uphole location. This method attempts to construct a model of the WL by estimating the velocity and thickness of the WL at several locations and interpolating between these locations. This method has the advantage of providing highly accurate near-surface velocities and thicknesses. However, it is very costly if high lateral resolution is required. Therefore, this method is used in estimating long-wavelength statics. Wavelength of statics refers to the width of the lateral (velocity or thickness) change in the WL relative to the spread length (maximum offset).

## 7.2.2   REFRACTION STATICS

This method is especially effective in estimating long-wavelength statics and is used to construct a model of the WL by estimating the velocity and thickness of the WL. The following are standard methods used for refraction statics calculation:

1. Delay-time Method: It uses the slopes of the direct and head waves as well as the head wave's intercept time of many shot records along the profile to estimate the WL velocity and thickness under each receiver. This method requires picking first breaks, which is difficult. It also requires reversed raypath[1] geometries, which might not be available.

2. Generalized reciprocal Method (GRM): It uses reversed refraction profiles to estimate the optimum WL thickness under each receiver. This method requires picking first breaks, which is difficult. It also requires reversed raypath geometries, which might not be available.

3. Least-squares Method: It uses least square analysis to find the best-fit WL velocity-thickness model to the first arrivals (i.e., direct and head waves). This method employs similar concepts to those used for the surface-consistent method (that we will discuss in the coming subsection), but uses refraction rather than reflection data.

[1]A raypath is a line that is perpendicular to the wavefront in isotropic media.

## 7.2.3    SURFACE-CONSISTENT RESIDUAL STATIC CORRECTION METHOD

This method is especially effective in estimating short-wavelength statics. The basic assumption of this method is that the static shifts are time delays that depend only on the source and receiver locations on the surface, not on raypaths in the subsurface. This assumption is valid only if all raypaths, regardless of source-receiver offset, are vertical in the near surface. The surface-consistent assumption is generally good because the WL usually has a much lower velocity than the sub-WL and refraction towards the normal at its base tends to make raypaths vertical.

The total residual static time shift on any trace can be expressed as:

$$T_{ijk} = R_i + S_j + G_k + M_k X_{ij}^2, \tag{7.2}$$

where $R_i$ is the residual static time shift associated with the $i^{th}$ receiver position ($i = 1, \ldots, I$, where $I$ is the number of receivers used in the survey), $S_j$ is the residual static time shift associated with the $j^{th}$ source position ($j = 1, \ldots, J$, where $J$ is the number of sources used in the survey), $G_k$ is the difference in two-way traveltime (due to structure) at a reference CMP and the traveltime at the $k^{th}$ CMP ($k = 1, \ldots, K$, where $K$ is number of CMPs covered in the survey), and $M_k X_{ij}^2$ is the residual NMO associated with the trace generated by the $j^{th}$ source and recorded by the $i^{th}$ receiver and it accounts for possible imperfect NMO correction due to using imperfect NMO velocities for the $k^{th}$ CMP. Generally, we have more equations than unknowns for typical seismic surveys. This is a typical least-squares problem. Our objective is to find those $R_i$, $S_j$, $G_k$, and $M_k$ that will minimize the error energy between the observed and calculated $T_{ijk}$ using model parameters in Equation 7.2:

$$E = \sum_{i=1}^{I} \sum_{j=1}^{J} \sum_{k=1}^{K} [R_i + S_j + G_k + M_k X_{ij}^2 - T_{ijk}]^2. \tag{7.3}$$

### 7.2.3.1    Surface-Consistent Residual Static Correction in Practice

The most widely used method is the pilot trace method [2], which consists of the following steps.

1. A CMP with good SNR ratio is gained and NMO-corrected using a preliminary velocity function.

2. The CMP gather is stacked to produce the first pilot trace.

3. Each individual trace in this CMP gather is crosscorrelated with the first pilot trace.

4. Time shifts $T_{ijk}'$, which correspond to maximum crosscorrelations, are picked.

5. Shift each original trace by its corresponding time shift $T_{ijk}'$.

6. A second pilot trace is constructed by stacking the shifted traces in the gather.

7. The second pilot trace is, in turn, crosscorrelated with the original traces in the gather and new time shifts $T_{ijk}$ are computed.

8. Shift each original trace by its corresponding new time shift $T_{ijk}$.

9. The process is performed this way on all CMP gathers moving to left and/or right from the starting (reference) CMP gather.

The following parameters are important when picking the time shifts in practice:

(a) Maximum allowable shift: which is the maximum shift allowed for crosscorrelations where a value between 30 and 40 ms is reasonable.

(b) Correlation window: where it should be chosen in an interval with the highest possible SNR ratio.

The following MATLAB script uses the M-function `scr_static.m` that implements the above steps and apply them to our CMP NMO corrected data, followed by stacking those statically corrected CMP gathers:

```
1  cmp_start=205;
2  cmp_end=255;
3  lags=20;
4  Dsort_static=scr_static(Dsort,Hsort,cmp_start,cmp_end,lags);
5  [Dstacked_static,t,cmp_num]=sstack(Dsort_static,Hsort);
6  save
7  Book_Seismic_Data_gain_bpf_sdecon_gain_sorted_nmo_corrected_static...
8   Dsort Dsort_static Hsort Dstacked_static cmp_num t
```

Note that we selected 40 ms (equivalent to 20 samples) for our data set as seen in the above MATLAB script. Also, we selected the whole trace (0-3s), which is the default in our code, for our data set. Figure 7.1 shows both the stacked data before and after applying the surface-consistent residual static correction method. Clearly, the data quality has improved after applying the correction where we notice the extension of the continuity to many of the layers.

# 7.3   USEFUL MATLAB FUNCTIONS

`scr_static.m`

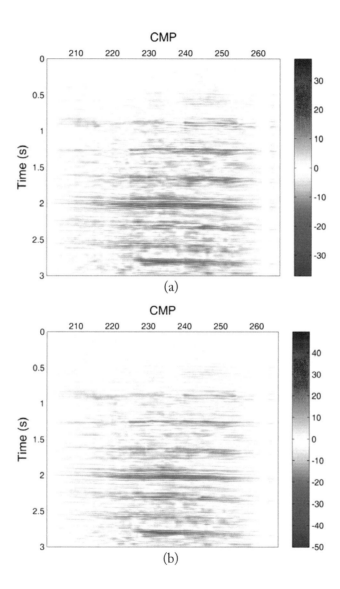

Figure 7.1: The stacked section of the east Texas seismic data set: (a) before applying residual static correction and (b) after applying residual static correction.

CHAPTER 8

# Seismic Migration

## 8.1 INTRODUCTION

After performing the necessary processing on the data and obtaining the stacked section, it can be very misleading to think of such data as a true image of the subsurface. This is because the actual reflection points are unknown. We need to take wave propagation effects into account to correctly determine the reflection points of the subsurface structure [26]. Improper correction of such undesirable geometrical effects leads to false interpretation and, therefore, oil/gas wells may be damaged or even misplaced. This is done using a process known as *Seismic Migration*. Hence, we can define migration as the process of reconstructing a seismic section so that the reflection events are repositioned under their correct surface locations at their correct vertical reflection time or depth location [1, 2]. Basically, migration removes the distorting effects of dipping reflectors from seismic sections. It also removes the diffracted arrivals resulting from sharp lateral discontinuities [27, 28, 29, 30]. There exist various migration techniques which are dependent upon the type of migration that suites our data or that we want to perform. In this chapter, we are going to perform post-stack time migration since our data mainly contains flat to gently-dipping layers as seen in its stacked section (Figure 7.1).

## 8.2 HUYGENS' PRINCIPLE AND BASIC MIGRATION PRINCIPLES

For this section, we will describe the physical principles of migration and we will follow [2, 31, 32]. Huygens' principle is the basis of migration. Consider the harbor example shown in Figure 8.1. Let us assume that a storm barrier exists at some distance $z_3$ from the beach and that there is a gap in the barrier. Now, also imagine that a calm afternoon breeze that comes from the ocean causes a plane incident water wave to hit the barrier. Its wavefront is parallel to the storm barrier. As we walk along the beach line, we notice a different wavefront from the incoming plane wave. The gap on the storm barrier has acted as a secondary source and generated the semicircular wavefront that is propagating towards the beach. Now, assume that we did not know about the storm barrier and the gap. We may want to lay out a receiver cable along the beach to record in time the approaching waves. Figure 8.2 illustrates this idea of recording in time the approaching waves with semicircular wavefronts. So the gap in the storm barrier acts as a Huygens' secondary source.

We can apply this principle to reflection seismology by imagining that each point on a reflector (geological interface) generates a secondary source in response to the incident wavefield. This is known as the *exploding reflector model* [31]. Consider a single point scatterer in a medium as shown

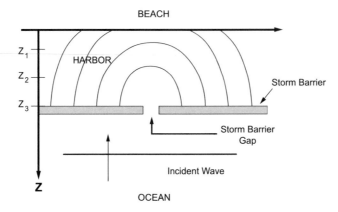

**Figure 8.1:** The beach example for illustrating Huygens' principle. The ocean causes a plane wave to hit the storm barrier where a different wavefront moving towards the beach is noticed after the storm barrier due to the barrier gap (modified after [2]).

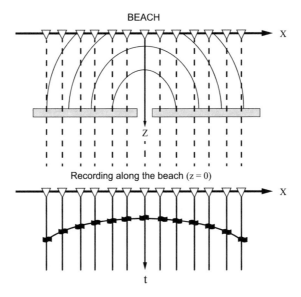

**Figure 8.2:** This figure shows how the approaching waves with semi-circular wavefronts generated by Huygens' secondary source from the beach example (Figure 8.1) are recorded as hyperbolas in time.

in Figure 8.3(a). The minimum travel time is given by:

$$t_0 = \frac{2z}{c}, \tag{8.1}$$

where $z$ is the depth of the scatter and $c$ is the velocity of the propagating wave (which we assume to be constant). Also, assume that the source and the receiver are co-located (zero-offset) as in

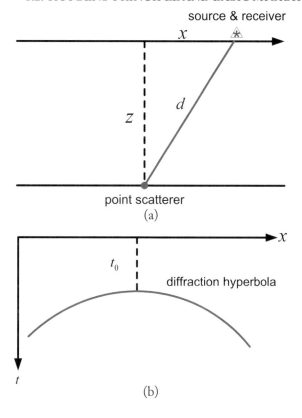

**Figure 8.3:** (a) A point scatterer (acting as Huygens' secondary source), and (b) a curved reflector which is produced based on the point scatterer.

Figure 8.3(a). The travel time as a function of distance, $x$, is given by [32]:

$$t(x) = \frac{2\sqrt{x^2 + z^2}}{c}. \tag{8.2}$$

By squaring, rearranging, and using Equation (8.1) then, Equation (8.2) can be expressed as:

$$\frac{t(x)^2}{t_0^2} - \frac{4x^2}{c^2 t_0^2} = 1. \tag{8.3}$$

This shows us that the travel time curve for the scattered arrival has the form of a hyperbola with the apex directly above the scattering point (our secondary source) as seen in Figure 8.3(b).

Now, consider a horizontal reflector (Figure 8.4(a)) that is composed of a series of point scatterers (gaps), each one of which generates a diffraction hyperbola in a zero-offset section as in Figure 8.4(b). Following Huygens' principle, these hyperbolas sum coherently only at the time of the reflection while the later contributions cancel out (Figure 8.4(b)). However, if the reflector vanishes at some point, then there will be a diffracted arrival from the endpoint that will show up in zero-offset data. This creates an artifact in the section that might be falsely interpreted as structure. We need to migrate such sections in order to remove such artifacts.

Another principle goal of migration is to map the apparent dip that is seen on the zero-offset sections into true dip [2, 26, 31, 33]. True dip angle is always greater than apparent dip angle. Consider a reflector dipping at an angle of $\theta$ in the true earth as in Figure 8.5. The zero-offset travel time for a wavefield propagating from distance $x$ down to the reflector and back up again is given by $t = 2r/c$ where $r$ is the wavefield path length and is equal to $r = x \sin \theta$. Now, to compare the apparent and true dip angles, the travel time must be converted to depth via Equation (8.1) and, therefore, in the unmigrated depth section $z = x \sin \theta$. By definition, the slope of this event is the tangent of the apparent dip angle, say $\beta$. Therefore, we have:

$$\tan \beta = \sin \theta. \tag{8.4}$$

Equation (8.4) shows clearly that the apparent dip angle is always less than true dip angle. Again in Figure 8.5 the events associated with the two zero-offset wavefields drawn from the dipping reflector to the two receivers will appear on the unmigrated section at the position locations associated with the two receivers [2]. Therefore, migration moves the energy up dip. In addition, from the same figure, the length of the reflector in the geological section is shorter than in the time section. Thus, migration also shortens reflectors. In summary, migration focuses energy by collapsing diffractions as well as it correctly steepens, shortens, and moves reflectors up-dip.

## 8.3   MIGRATION KINDS

Migration can be implemented based on different techniques. It is an important and expensive process that is applied to reflection seismic data before it is interpreted. Since it is the last major process applied to the data, it is likely to be blamed for all sorts of things like inconsistent amplitudes and lack of structural details even though these problems may arise from acquisition or earlier processing steps [34]. Therefore, it is important to know what type of migration to use.

Migration can be classified as *Pre-stack* migration or *Post-stack* migration. For the former, migration is performed on pre-stacked data either on shot gathers or on CMPs. For large surveys, it will require massive computer storage and days, maybe even weeks, of CPU time on a super computer [34]. For the latter case, the migration is applied on the stacked CMP data. It is much less expensive than pre-stack migration but it is also less accurate in complicated areas of the subsurface. Finally, it can be accomplished on workstation class machines.

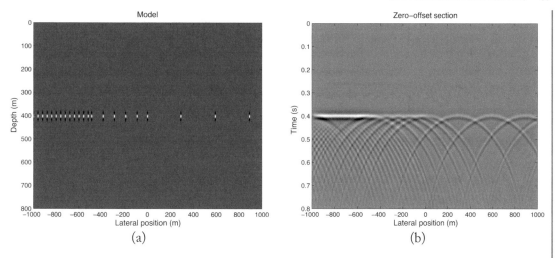

Figure 8.4: (a) An array of point scatterers positioned at different locations, and (b) its resultant curved reflectors interfering with each other.

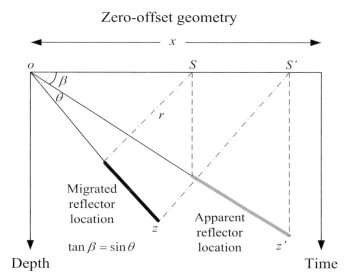

Figure 8.5: Migration principles. The apparent reflector with a dip angle $\beta$ in the time section when migrated is moved up-dip, steepened (to an angle $\theta$), shortened, and mapped on its true subsurface location.

Moreover, migration can be classified in terms of how much physics we put into the algorithm, specifically for handling velocity variations. This type of classification is based on time migration versus depth migration (see Figure 8.6). Note that any migration (pre-stack or post-stack) can be output in time or depth. In areas of strong velocity variations, depth migration is used and the output is given as a depth section. Geological examples of strong lateral velocity variations include salt over-hangs, sub-salt areas, or combinations of such features [34]. The main difference between time and depth migration is mostly for ease of interpretation afterwards where one can make more simplifica-tions in time migration than for depth migration. Time migration is valid only when lateral velocity variations are mild (10%) to moderate (30%). When this assumption fails, we have to use depth migration. Figure 8.7 illustrates schematically were pre/post stack time/depth migration is employed.

Finally, we also can classify migration into two-dimensional (2-D) and three-dimensional (3-D). In 2-D migration, we migrate the data once along the profile. This might generate miss-ties on intersecting profiles. In addition, 2-D migration is prone to sideswipe effects. Sideswipes are reflections from out of the plane of the profile. In contrast, 3-D migration first migrates the data in the inline direction then takes that migrated data and migrates it again in the crossline direction. This is the two-pass 3-D migration. A one-pass 3-D migration can also be done using a downward continuation approach. Therefore, considering 2-D versus 3-D, pre-stack versus post-stack, and time versus depth, we can have the following types of migrations (ordered from fastest but least accurate to slowest but most accurate) [35]:

1. 2-D poststack time migration (fastest, least-accurate)

2. 2-D poststack depth migration

3. 2-D prestack time migration

4. 2-D prestack depth migration

5. 3-D poststack time migration

6. 3-D poststack depth migration

7. 3-D prestack time migration

8. 3-D prestack depth migration (slowest, most accurate)

Since our data set is 2-D and contains mainly flat layers, we, therefore, choose to perform post-stack time migration.

## 8.4  MIGRATION ALGORITHMS

There exist many migration algorithms but mostly belong to the following three main post-stack migration algorithms:

1. Kirchhoff migration.

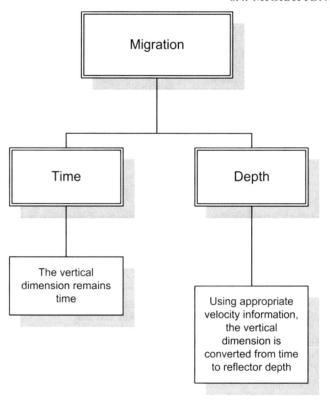

**Figure 8.6:** This is another migration classification which depends on how much physics one puts into the migration algorithm.

2. Frequency-wavenumber-$(f - k)$ migration.

3. Finite-difference (downward continuation) migration.

### 8.4.1   POST-STACK TIME MIGRATION USING THE STOLT $f - k$ MIGRATION

The zero-offset stack section requires moving the reflectors to their true positions and we can obtain this via post-stack time migration. We select to use the popular method of Stolt [2, 36] to do so. This method basically depends on one forward and one backward 2-D Fast Fourier Transform (FFT) computations as will as one mapping and a multiplication. Assuming that our stacked data is denoted by $u(t, x)$ (where $t$ represents time and $x$ is the distance), the algorithm works as follows [2]:

1. Take the 2-D Fourier transform of $u(t, x)$ to obtain $U(\omega, k_x)$, where $\omega$ is the angular frequency and $k_x$ is spatial wavenumber.

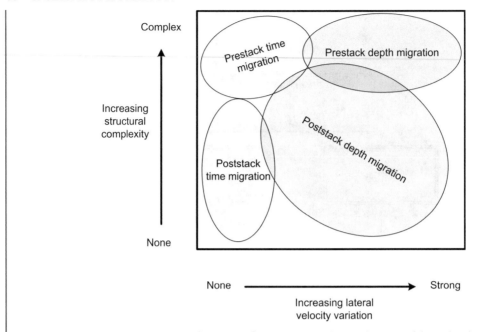

**Figure 8.7:** Migration types as a function of computational complexity and lateral velocity variations (after [34]). The most accurate and most expensive migration kind is the pre-stack depth migration.

2. Apply the following mapping to the $\omega$ variable to become the depth wavenumber $k_z$ where we map $U(\omega, k_x)$ to become $U(k_z, k_x)$:

$$k_z = \sqrt{(\frac{2\omega}{v})^2 - k_x^2}.$$

3. Then we compute a scale value:

$$S = \frac{v}{2} \frac{k_z}{\sqrt{k_z^2 + k_x^2}}$$

and multiply it by $U(k_z, k_x)$.

4. Finally, take the 2-D inverse Fourier transform of the result and this will be our migrated section.

The following script uses the M-function `stolt_mig.m` to perform post-stack time migration based on Stolt's $f - k$ migration algorithm [36]:

```
1
2  load
3  Book_Seismic_Data_gain_bpf_sdecon_gain_sorted_nmo_corrected_static...
4  load Book_Seismic_Data_gain_bpf_sdecon_gain_sorted_velocities.mat
5
6  dx=110;
7  dt=0.002;
8  v=mean(mean(v_stack));
9  Dstacked_static=Dstacked_static(:,1:60);
10 cmp_num=cmp_num(1:60)
11
12 Dmigrated=stolt_mig(Dstacked,v,dt,dx);
13
14 figure,simage_display(Dstacked_static,cmp_num,t,0);
15 xlabel('CMP','FontSize',14)
16 ylabel('Time (s)','FontSize',14)
17 title('Stacked section','FontSize',14)
18
19 figure,simage_display(Dmigrated,cmp_num,t,0);
20 xlabel('CMP','FontSize',14)
21 ylabel('Time (s)','FontSize',14)
22 title('Migrated section','FontSize',14)
```

Note that when migrating the data we have used the first 60 traces (out of 65) of our stacked data since trace 61 is equal to zero (refer to Figure 7.1. Figure 8.8 shows the comparison between the stacked and post-stack time migrated section based on Stolt's $f - k$ algorithm. The horizontal resolution was improved after we applied the post-stack time migration algorithm to our stacked section as we can see, for example, at time equal to 2, 2.5 and 2.75 s. That is, we have improved the continuity of the layers. Of course, the migration result will be improved if we apply post-stack depth migration but this requires estimating the velocity model for this geological section and then performing the post-stack depth migration in the $f - k$ domain as described in [2].

## 8.5   USEFUL MATLAB FUNCTIONS

stolt_mig.m.

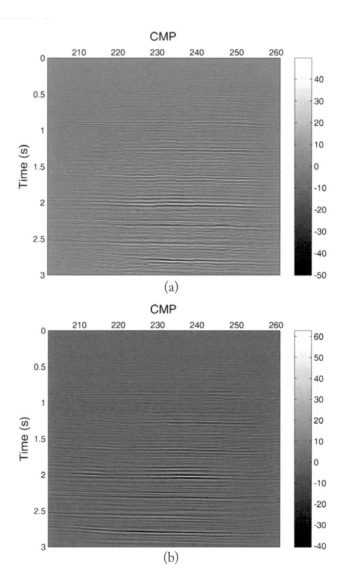

**Figure 8.8:** The east Texas seismic data set: (a) its stacked section and (b) its post-stack migrated section using Stolt's $f - k$ migration algorithm.

CHAPTER 9

# Concluding Remarks

In this book, we introduced the basic concepts encountered in a typical seismic data processing flow. The concepts were then applied on a real 2-D data set from east Texas, USA. The processes applied to the data set included:

1. Obtaining geometry information

2. Quality control of the data

3. Ground-roll attenuation using band-pass frequency filtering

4. Spiking deconvolution using Wiener optimum deconvolution

5. CMP sorting

6. Velocity analysis using the velocity spectrum method

7. NMO correction and stretch muting

8. Residual static correction using the surface-consistent method

9. Stacking

10. Migration using the $f - k$ (Stolt) migration

MATLAB codes were used to perform each of these processes. Most of these codes were written by the authors while few others were borrowed from public-domain resources. We made genuine attempts to make these codes as simple as possible with plenty of comments to encourage interested readers to modify them to fit their own data sets. Nevertheless, we greatly appreciate feedback from our readers.

# Bibliography

[1] P. Kearey, M. Brooks, and I. Hill. *An Introduction to Geophysical Exploration*. Blackwell Science, 3rd edition, 2002. Cited on page(s) 1, 2, 24, 61

[2] Öz. Yilmaz, editor. *Seismic Data Analysis: Processing, Inversion, and Interpretation of Seismic Data*. Society of Exploration Geophysicists, 2nd edition, 2001. Cited on page(s) 1, 4, 7, 17, 23, 24, 30, 34, 35, 39, 48, 55, 57, 61, 62, 64, 67, 69

[3] A. S. Spanias, S. B. Jonsson, and S. D. Stearns. Transform methods for seismic data compression. *IEEE Tran. on Geosciences and Remote Sensing*, 29(3):407–416, May 1991. DOI: 10.1109/36.79431 Cited on page(s) 1, 23

[4] B. Buttkus. *Spectral Analysis and Filter Theory in Applied Geophysics*. Springer, 2000. Cited on page(s) 1, 24, 33

[5] D. Forel, T. Benz, and W. D. Pennington. *Seismic Data Processing with Seismic Un*x: A 2-D Seismic Data Processing Primer*. Society of Exploration Geophysicists (SEG), 2005. Cited on page(s) 7

[6] J. F. Clearbout. *Imaging the Earth's interior*. Blackwell Scientific Publications, 1985. Cited on page(s) 20

[7] X. Miao and S. Cheadle. Noise attenuation with wavelet transform. *SEG 1998 Expanded Abstracts*, 1998. DOI: 10.1190/1.1820071 Cited on page(s) 23

[8] T. J. Ulrych, M. D. Sacchi, and J. M Graul. Signal and noise separation: Art and science. *Geophysics*, 64:1648 – 1656, Sept.- Oct. 1999. DOI: 10.1190/1.1444670 Cited on page(s) 23

[9] A. Özbek. Adaptive beamforming with generalized linear constrains. *Geophysics Extended Abstracts*, 2000. DOI: 10.1190/1.1815855 Cited on page(s) 23

[10] L. Duval and T. Rosten. Filter bank decomposition of seismic data with application to compression and denoising. *SEG 2000 Expanded Abstracts*, 2000. DOI: 10.1190/1.1815847 Cited on page(s) 23

[11] Qiansheng Cheng, Rong Chen, and Ta-Hsin Li. Simultaneous wavelet estimation and deconvolution of reflection seismic signals. *IEEE Transactions on Geoscience and Remote Sensing*, 34(2):377 – 384, March 2001. DOI: 10.1109/36.485115 Cited on page(s) 23

[12] K. Berkner and Jr. Wells, R.O. Wavelet transforms and denoising algorithms. *Conference Record of the Thirty-Second Asilomar Conference on Signals, Systems and Computers*, 2:1639 – 1643, Nov. 1998. Cited on page(s) 23

[13] Rongfeng Zhang and Tadeusz J. Ulrych. Physical wavelet frame denoising. *Geophysics*, 68(1):225–231, Jan 2003. DOI: 10.1190/1.1543209 Cited on page(s) 23

[14] J. E. Womack and J. R. Cruz. Seismic data filtering using a gabor representation. *IEEE Tran. on Geosciences and Remote Sensing*, 32(2):467–472, March 1994. DOI: 10.1109/36.295061 Cited on page(s) 24

[15] A. F. Linville and R. A. Meek. A procedure for optimally removing localized coherent noise. *Geophysics*, 60(1):191 – 203, Jan.-Feb. 1995. DOI: 10.1190/1.1443746 Cited on page(s) 24

[16] B. Duquet and K. J. Marfurt. Filtering coherent noise during prestack depth migration. *Geophysics*, 64(4):1054 – 1066, July-Aug. 1999. DOI: 10.1190/1.1444613 Cited on page(s) 24

[17] S. Treitel, J. L. Shanks, and C. W. Fraster. Some aspects of fan filtering. *Geophysics*, XXXII:789 – 800, 1967. DOI: 10.1190/1.1439889 Cited on page(s) 24

[18] M. Z. Mulk, K. Obata, and K. Hirano. Design of fan filters. *IEEE Trans. on Acoustics, Speech, and Signal Processing*, 31(6):1427 – 1434, Dec. 1983. Cited on page(s) 24

[19] A. H. Kayran and R. A. King. Design of recursive and nonrecursive fan filters with complex transformation. *IEEE Trans. on Circuits and Systems*, 30(12):849 – 857, Dec. 1983. DOI: 10.1109/TCS.1983.1085321 Cited on page(s) 24

[20] D. W. McCowan, P. L. Stoffa, and J. B. Diebold. Fan filters for data with variable spatial sampling. *IEEE Trans. on Acoustics, Speech, and Signal Processing*, 32(6):1154 – 1159, Dec. 1984. DOI: 10.1109/TASSP.1984.1164466 Cited on page(s) 24

[21] R. Ansari. Efficient IIR and FIR fan filters. *IEEE Trans. on Circuits and Systems*, 34:941 – 945, August 1987. DOI: 10.1109/TCS.1987.1086224 Cited on page(s) 24

[22] R. H. Bamberger and M. J. T Smith. A filter bank for the directional decomposition of images: Theory and design. *IEEE Tran. on Signal Processing*, 40(4):882–893, December 1992. DOI: 10.1109/78.127960 Cited on page(s) 24

[23] E. A. Robinson and S. Treitel. *Geophysical Signal Analysis*. SEG, 2000. Cited on page(s) 33, 34

[24] E. A. Robinson and S. Treitel. *Digital Imaging & Deconvolution: The ABCs of seismic exploration and processing*. SEG, 2008. Cited on page(s) 33

[25] N. Neidell and M. T. Tanner. Semblance and other co-herency measures for multichannel data. *Geophysics*, 36:482–497, 1971. DOI: 10.1190/1.1440186 Cited on page(s) 47

[26] E. A. Robinson. *Migration of geophysical data.* Intr. Human Resources Development Corporation, 1983. Cited on page(s) 61, 64

[27] J. W. Thorbeck and A. J. Berkhout. 3-D recursive extrapolation operators: an overview. *Geophysics Extended Abstracts*, 1994. Cited on page(s) 61

[28] V. K. Madisetti and D. B. Williams, editors. *The Digital Signal Processing Handbook.* CRC Press and IEEE Press, 1998. Cited on page(s) 61

[29] D. E. Dudgeon and R. M. Mersereau. *Multidimensional Digital Signal Processing.* Prentice-Hall, 1984. Cited on page(s) 61

[30] L. J. Karam and J. H. McClellan. Efficient design of digital filters for 2-D and 3-D depth migration. *Signal Processing, IEEE Transactions on*, 45(4):1036–1044, April 1997. DOI: 10.1109/78.564191 Cited on page(s) 61

[31] J. F. Claerbout. *Imaging the Earth's Interior.* Blackwell, 1984. Cited on page(s) 61, 64

[32] P. M. Shearer. *Introduction to Seismology.* Cambridge Uni. Press, 1999. Cited on page(s) 61, 63

[33] J. A. Scales. *Theory of Seismic Imaging.* Samizdat Press, 1997. Cited on page(s) 64

[34] C. L. Liner. *Elements of 3-D Seismology.* PennWell, 1999. Cited on page(s) 64, 66, 68

[35] B. L. Biondi. *3D Seismic Imaging.* SEG, 2006. Cited on page(s) 66

[36] R. H. Stolt. Migration by Fourier transform. *Geophysics*, 43(1):23–48, 1978. DOI: 10.1190/1.1440826 Cited on page(s) 67, 68

# Authors' Biographies

## WAIL A. MOUSA

**Wail A. Mousa** holds two B.S. with honors in electrical engineering and mathematical sciences and an M.S. in electrical engineering from KFUPM. Wail was the first Saudi to be sponsored by Schlumberger Dhahran Carbonate Research (SDCR) for a Ph.D., which he obtained at the School of Electronics & Electrical Engineering at the University of Leeds in 2006. He was a graduate assistant at KFUPM and then promoted to lecturer between 2000-2003. Wail then worked between 2003-2009 as a research scientist on applied signal processing for geology and geophysics at the SDCR. He is currently working as an assistant professor in the Electrical Engineering Department, King Fahd University of Petroleum & Minerals (KFUPM), teaching various signal and image processing courses and supervising students. Wail is an active researcher and publishes regularly where his research interests include: digital signal, image, and video processing and their applications in geophysics, geology as well as petro-physics, design & implementation of digital filters including wavefield extrapolation filters, image segmentation, pattern recognition and classification and their applications related to geophysical and geological data, signal and image compression. More recently, he published his first patent in the US patent and trade office. He was additionally appointed as the director of business development at KFUPM as of February 2010. He is a member of many international and local professional organizations. In late 2006, he represented both Saudi Arabia in the inaugural World Petroleum Council-Youth Committee (WPC-YC) and in early 2007, he became the vice chair of the WPC-YC, where he has been chairman since June 2008. Dr. Mousa has received many awards of special honor including a letter of distinction from the CEO of Schlumberger in 2008 due to his technical contributions to the company since he joined them after his PhD.

## ABDULLATIF A. AL-SHUHAIL

**Abdullatif A. Al-Shuhail** is an associate professor of Geophysics at King Fahd University of Petroleum & Minerals (KFUPM). He received his B.S. degree in Geophysics from KFUPM in 1987. He got his M.Sc. and Ph.D. in Geophysics from Texas A&M University, College Station. He founded the Near Surface Seismic Investigation Consortium at KFUPM in 2006. His fields of interests include seismic characterization of fractured reservoirs, near-surface effects on petroleum seismic data, and ground penetrating radar. He is member of the Society of Exploration Geophysicists (SEG), European Association of Geoscientists and Engineers (EAGE), and Dhahran Geoscience Society (DGS).

# Index

Printed in the United States
by Baker & Taylor Publisher Services